Microcontroller Projects in C for the 8051

Microcontroller Projects in C for the 8051

Dogan Ibrahim

Newnes

AMSTERDAM BOSTON HEIDELBERG LONDON NEW YORK OXFORD
PARIS SAN DIEGO SAN FRANCISCO SINGAPORE SYDNEY TOKYO

Newnes
An imprint of Elsevier
Lineacre House, Jordan Hill, Oxford OX2 8DP
200 Wheeler Road, Burlington, MA 01803

First published 2000
Reprinted 2003

Copyright © 2000, Dogan Ibrahim. All rights reserved

The right of Dogan Ibrahim to be identified as the author of
this work has been asserted in accordance with the Copyright,
Designs and Patents Act 1988.

No part of this publication may be reproduced in any material form
(including photocopying or storing in any medium by electronic means
and whether or not transiently or incidentally to some other use of this
publication) without the written permission of the copyright holder except
in accordance with the provisions of the Copyright, Designs and Patents
Act 1988 or under the terms of a licence issued by the Copyright Licensing
Agency Ltd, 90 Tottenham Court Road, London, England W1T 4LP.
Applications for the copyright holder's written permission to reproduce
any part of this publication should be addressed to the publisher.

Permissions may be sought directly from Elsevier's Science and Technology
Rights Department in Oxford, UK; phone: (+44) (0) 1865 843830;
fax: (+44) (0) 1865 853333; e-mail: permissions@elsevier.co.uk. You may also
complete your request on-line via the Elsevier homepage
(http://www.elsevier.com), by selecting 'Customer Support' and then
'Obtaining 27Permissions'.

British Library Cataloguing in Publication Data
A catalogue record for this book is available from the British Library

Library of Congress Cataloging in Publication Data
A catalogue record for this book is available from the Library of Congress

ISBN 0 7506 4640 3

> For information on all Newnes publications
> visit our website at www.newnespress.com

Typeset by David Gregson Associates, Beccles, Suffolk
Printed and bound by Antony Rowe Ltd, Eastbourne

Contents

Preface — vii

Chapter 1   Microcomputer Systems — 1
   1.1   Introduction — 1
   1.2   Microcontroller Evolution — 1
   1.3   Microcontroller Architecture — 2
   1.4   8051 Family — 3
   1.5   Architecture of the 8051 Family — 4
   1.6   Pin Configuration — 4
   1.7   Timer/Counters — 10
   1.8   Interrupt Control — 11
   1.9   Minimum Microcontroller Configuration — 12
   1.10 Project Development — 13

Chapter 2   Programming Microcontrollers in C — 15
   2.1   Data Types — 16
      2.1.1   bit — 16
      2.1.2   signed char/unsigned char — 16
      2.1.3   signed short/unsigned short — 17
      2.1.4   signed int/unsigned int — 17
      2.1.5   signed long/unsigned long — 18
      2.1.6   float — 18
      2.1.7   sbit — 18
      2.1.8   sfr — 19
      2.1.9   sfr16 — 19
   2.2   Memory Models — 19
   2.3   Interrupts — 20
   2.4   Structure of a Microcontroller-based C Program — 21
   2.5   Program Description Language — 22
      2.5.1   START-END — 22
      2.5.2   Sequencing — 24
      2.5.3   IF-THEN-ELSE-ENDIF — 24

2.5.4 DO-ENDO	24
2.5.5 REPEAT-UNTIL	25
2.6 Internet Web Sites of Microcontroller Compilers	25
2.7 Further Reading	27

Chapter 3 Light Projects — 29

PROJECT 1 – LED Binary Counter	29
PROJECT 2 – LED Chasing Circuit	33
PROJECT 3 – Random LED Pattern	34
PROJECT 4 – Cyclic LED Pattern	37
PROJECT 5 – LED Dice	38
PROJECT 6 – Hexadecimal Display	46
PROJECT 7 – Two-digit Decimal Count	50
PROJECT 8 – TIL311 Dice	53
PROJECT 9 – 7 Segment Display Driver	57
PROJECT 10 – Four-digit LED Display Interface	62
PROJECT 11 – Interrupt Driven Event Counter with 4-digit LED Display	75

Chapter 4 Sound Projects — 85

PROJECT 12 – Simple Buzzer Interface	86
PROJECT 13 – Small Speaker Interface (Using the Timer Interrupt)	90
PROJECT 14 – Two-tone Small Speaker Interface (Using the Timer Interrupt)	94
PROJECT 15 – Electronic Siren (Using the Timer Interrupt)	95
PROJECT 16 – Electronic Siren (Using the Timer Interrupt)	101

Chapter 5 Temperature Projects — 107

PROJECT 17 – Using a Digital Temperature Sensor	108
PROJECT 18 – Digital Thermometer with Centigrade/Fahrenheit Output	119
PROJECT 19 – Digital Thermometer with High Alarm Output	125
PROJECT 20 – Digital Thermometer with High and Low Alarm Outputs	126
PROJECT 21 – Using Analogue Temperature Sensor IC with A/D Converter	132

Chapter 6 RS232 Serial Communication Projects — 147

PROJECT 22 – Output a Simple Text Message from the RS232 Port	151
PROJECT 23 – Input/Output Example Using the RS232 Port	155
PROJECT 24 – A Simple Calculator Program Using the RS232 Port	161

Appendix – ASCII code	167
GLOSSARY	171
Index	177

Preface

A microcontroller is a single chip microprocessor system which contains data and program memory, serial and parallel I/O, timers, external and internal interrupts, all integrated into a single chip that can be purchased for as little as $2.00. It is estimated that on average, a middle-class household in America has a minimum of 35 microcontrollers in it. About 34% of microcontroller applications are in office automation, such as laser printers, fax machines, intelligent telephones, and so forth. About one-third of microcontrollers are found in consumer electronics goods. Products like CD players, hi-fi equipment, video games, washing machines, cookers and so on fit into this category. The communications market, automotive market, and the military share the rest of the application areas.

Microcontrollers have traditionally been programmed using the assembly language of the target microcontroller. Different microcontrollers from different manufacturers have different assembly languages. Assembly language consists of short mnemonic descriptions of the instruction sets. These mnemonics are difficult to remember and the programs developed for one microcontroller cannot be used for other types of microcontrollers. The most common complaint about microcontroller programming is that the assembly language is somewhat difficult to work with, especially during the development of complex projects. The solution to this problem is to use high-level languages. This makes the programming a much simpler task and the programs are usually more readable, portable, and easier to maintain. There are various forms of BASIC and C compilers available for most microcontrollers. BASIC compilers are usually in the form of interpreters and the code produced is usually slow.

Another disadvantage of BASIC is that most BASIC compilers are not structured and this makes the program maintenance a difficult task. In this book we shall be using a fully featured professional C compiler to program our target microcontroller.

This book is about programming the 8051 family of microcontrollers using the C language, and I have chosen the AT89C2051 microcontroller for all the examples. AT89C2051 belongs to the industry standard 8051 family of microcontrollers. AT89C2051 is a 20-pin device which is fully code compatible with its bigger brother 8051. The device contains a serial port, 15 bits parallel I/O, two timer/counters, six interrupt sources, 128 bytes of data RAM, and 2 Kbytes of reprogrammable flash program memory. There are many reasons for choosing the AT89C2051, including its compatibility with the 8051 family and the ease of erasing and reprogramming the device. There is no need to use a UV eraser to erase the program memory. The memory can be erased and then reprogrammed by using a low-cost programmer. Other reasons for using the AT89C2051 are its low cost and small size. All of the examples given herein can run on all members of the 8051 family.

Chapter 1 provides an introduction to the architecture of the 8051 family, with special emphasis on the AT89C2051 microcontroller. Chapter 2 describes the features of the C compiler used throughout the projects in this book. Addresses of some popular web sites are also given in this chapter which contain information on the 8051 family. Chapter 3 provides many light-based projects. The circuit diagrams and the full C code of all the projects are given with full comments and explanations. All the projects have been built and tested on a breadboard. Chapter 4 is based on sound projects and there are working projects from simple buzzer circuits to electronic organ projects. Chapter 5 provides several working temperature-based projects using digital temperature sensors and analogue-to-digital converters. Finally, Chapter 6 describes several RS232-based projects which explain how information can be transferred between a microcontroller and external devices.

Dogan Ibrahim
1999, London

CHAPTER 1

MICROCOMPUTER SYSTEMS

1.1 Introduction

The term microcomputer is used to describe a system that includes a microprocessor, program memory, data memory, and an input/output (I/O). Some microcomputer systems include additional components such as timers, counters, analogue-to-digital converters and so on. Thus, a microcomputer system can be anything from a large computer system having hard disks, floppy disks and printers, to single chip computer systems.

In this book we are going to consider only the type of microcomputers that consist of a single silicon chip. Such microcomputer systems are also called microcontrollers.

1.2 Microcontroller Evolution

First, microcontrollers were developed in the mid-1970s. These were basically calculator-based processors with small ROM program memories, very limited RAM data memories, and a handful of input/output ports.

As silicon technology developed, more powerful, 8-bit microcontrollers were produced. In addition to their improved instruction sets, these microcontrollers included on-chip counter/timers, interrupt facilities, and improved I/O handling. On-chip memory capacity was still small and was not adequate for many applications. One of the most significant developments at this time was the availability of on-chip ultraviolet erasable EPROM memory. This simplified the product development time considerably and, for the first time, also allowed the use of microcontrollers in low-volume applications.

The 8051 family was introduced in the early 1980s by Intel. Since its introduction, the 8051 has been one of the most popular microcontrollers and has been second-sourced by many manufacturers. The 8051 currently has many different versions and some types include on-chip analogue-to-digital converters, a considerably large size of program and data memories, pulse-width modulation on outputs, and flash memories that can be erased and reprogrammed by electrical signals.

Microcontrollers have now moved into the 16-bit market. 16-bit microcontrollers are high-performance processors that find applications in real-time and compute intensive fields (e.g. in digital signal processing or real-time control). Some of the 16-bit microcontrollers include large amounts of program and data memories, multi-channel analogue-to-digital converters, a large number of I/O ports, several serial ports, high-speed arithmetic and logic operations, and a powerful instruction set with signal processing capabilities.

1.3 Microcontroller Architecture

The simplest microcontroller architecture consists of a microprocessor, memory, and input/output. The microprocessor consists of a central processing unit (CPU) and the control unit (CU).

The CPU is the brain of a microprocessor and is where all of the arithmetic and logical operations are performed. The control unit controls the internal operations of the microprocessor and sends control signals to other parts of the microprocessor to carry out the required instructions.

Memory is an important part of a microcomputer system. Depending upon the application we can classify memories into two groups: program memory and data memory. Program memory stores all the program code. This memory is usually a read-only memory (ROM). Other types of memories, e.g. EPROM and PEROM flash memories, are used for low-volume applications and also during program development. Data memory is a read/write memory (RAM). In complex applications where there may be need for large amounts of memory it is possible to interface external memory chips to most microcontrollers.

Input/Output (I/O) ports allow external digital signals to be connected to the microcontroller. I/O ports are usually organized into groups of 8 bits and each group is given a name. For example, the 8051 microcontroller contains four 8-bit I/O ports named P0, P1, P2, and P3. On some microcontrollers the direction of the I/O port lines are programmable so that different bits of a port can be programmed as inputs or outputs. Some microcontrollers (including the 8051 family) provide bi-directional I/O ports. Each I/O port line of such microcontrollers can be used as inputs and outputs. Some microcontrollers provide 'open-drain' outputs where the output transistors are left floating (e.g. port P0 of the 8051 family). External pull-up resistors are normally used with such output port lines.

1.4 8051 Family

The 8051 family is a popular, industry standard 8-bit single chip microcomputer (microcontroller) family, manufactured by various companies with many different capabilities. The basic standard device, which is the first member of the family, is the 8051, which is a 40-pin microcontroller. This basic device is now available in several configurations. The 80C51 is the low-power CMOS version of the family. The 8751 contains EPROM program memory, used mainly during development work. The 89C51 contains flash programmable and erasable memory (PEROM) where the program memory can be reprogrammed without erasing the chip with ultraviolet light. The 8052 is an enhanced member of the family which contains more RAM and also more timer/counters. There are many versions of the 40-pin family which contain on-chip analogue-to-digital converters, pulse-width modulators, and so on. At the lower end of the 8051 family we have the 20-pin microcontrollers which are code compatible with the 40-pin devices. The 20-pin devices have been manufactured for less complex applications where the I/O requirements are not very high and where less power is required (e.g. in portable applications). The AT89C1051 and AT89C2051 (manufactured by Atmel) are such microcontrollers, which are fully code compatible with the 8051 family and offer reduced power and less functionality. Table 1.1 gives a list of the characteristics of some members of the 8051 family.

Table 1.1 Some popular members of the 8051 family

Device	Program memory	Data memory	Timer/ counters	I/O pins	Pin count
AT89C1051	1K flash	64 RAM	1	15	20
AT89C2051	2K flash	128 RAM	2	15	20
AT89C51	4K flash	128 RAM	2	32	40
AT89C52	8K flash	256 RAM	3	32	40
8051AH	4K ROM	128 RAM	2	32	40
87C51H	4K EPROM	128 RAM	2	32	40
8052AH	8K ROM	256 RAM	3	32	40
87C52	8K EPROM	256 RAM	3	32	40
87C54	16K EPROM	256 RAM	3	32	40
87C58	32K EPROM	256 RAM	3	32	40

In this book all the projects are based upon the AT89C2051 microcontroller. The code given will run on other members of the family, including the 40-pin devices. The reasons for choosing the AT89C2051 are its low cost, low power consumption, small space (20 pin), and powerful features.

In this chapter we shall be looking at the features of the 8051 family briefly with more emphasis on the smaller AT89C2051. More information on these microcontrollers can be obtained from the manufacturers' data sheets.

1.5 Architecture of the 8051 Family

The 8051 is an 8-bit, low-power, high-performance microcontroller. There are a large number of devices in the 8051 family with similar architecture and each member of the family is downward compatible with each other. The basic 8051 microcontroller has the following features:

- 4 Kbytes of program memory
- 256×8 RAM data memory
- 32 programmable I/O lines
- Two 16-bit timer/counters
- Six interrupt sources
- Programmable serial UART port
- External memory interface
- Standard 40-pin package

The EPROM versions of the family (e.g. 8751) are used for development and the program memory of these devices is erased with an ultraviolet light source. The pin configuration of the standard 8051 microcontroller is shown in Fig. 1.1.

The AT89C2051 is a low-end member of the 8051 family, aimed for less complex applications. This device contains a 2 Kbyte flash programmable memory (PEROM) which can be erased and reprogrammed using a suitable programmer. The AT89C2051 contains 128 bytes of RAM and 15 programmable I/O lines. The code developed for this device runs on a standard 8051 without any modification. As shown in Fig. 1.2, the AT89C2051 is housed in a 20-pin package.

1.6 Pin Configuration

Descriptions of the various pins are given below.

Microcomputer Systems 5

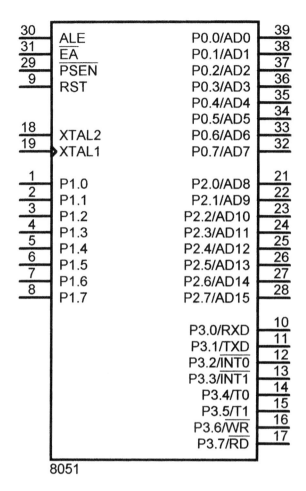

Figure 1.1.
Pin configuration of the standard 8051

RST

This is the reset input. This input should normally be at logic 0. A reset is accomplished by holding the RST pin high for at least two machine cycles. Power-on reset is normally performed by connecting an external capacitor and a resistor to this pin (see Figs 1.3 and 1.4).

P3.0

This is a bi-directional I/O pin (bit 0 of port 3) with an internal pull-up resistor. This pin also acts as the data receive input (RXD) when the device is used as an asynchronous UART to receive serial data.

6 Microcontroller Projects in C for the 8051

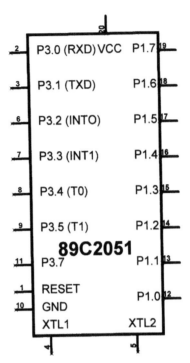

Figure 1.2.
Pin configuration of the standard AT89C2051

P3.1

This is a bi-directional I/O pin (bit 1 of port 3) with an internal pull-up resistor. This pin also acts as the data transmit output (TXD) on the 8051 when the device is used as an asynchronous UART to transmit serial data.

XTAL1 and XTAL2

These pins are where an external crystal should be connected for the operation of the internal oscillator. Normally two 33 pF capacitors are connected with the crystal as shown in Figs 1.3 and 1.4. A machine cycle is obtained by dividing the crystal frequency by 12. Thus, with a 12 MHz crystal, the machine cycle is 1 μs. Most machine instructions execute in one machine cycle.

P3.2

This is a bi-directional I/O pin (bit 2 of port 3) with an internal pull-up resistor. This pin is also the external interrupt 0 (INT0) pin.

Microcomputer Systems

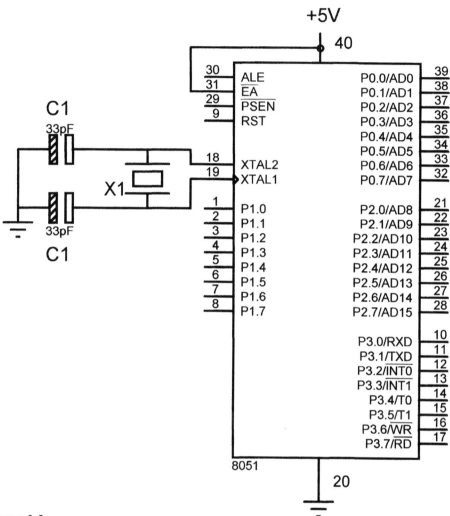

Figure 1.3.
Minimum 8051 configuration

P3.3

This is a bi-directional I/O pin (bit 3 of port 3) with an internal pull-up resistor. This pin is also the external interrupt 1 (INT1) pin.

P3.4

This is a bi-directional I/O pin (bit 4 of port 3) with an internal pull-up resistor. This pin is also the counter 0 input (T0) pin.

8   Microcontroller Projects in C for the 8051

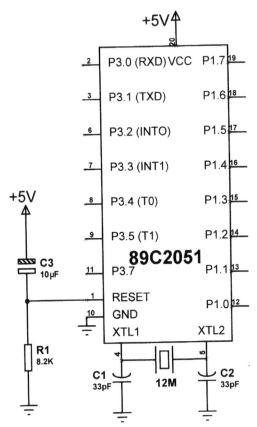

Figure 1.4.
Minimum AT89C2051 configuration

P3.5

This is a bi-directional I/O pin (bit 5 of port 3) with an internal pull-up resistor. This pin is also the counter 1 input (T1) pin.

GND

Ground pin.

P3.6

This is a bi-directional I/O pin. This pin is not available on the AT89C2051. It is also the external memory write (WR) pin.

P3.7

This is a bi-directional I/O pin for bit 7 of port 3. On the standard 8051, this pin is also the external data memory read (RD) pin.

P1.0

This is a bi-directional I/O pin for bit 0 of port 1. This pin has no internal pull-up resistors on the 20-pin devices. It is also used as the positive input of the analogue comparator (AIN0) on the 20-pin device.

P1.1

This is a bi-directional I/O pin for bit 1 of port 1. This pin has no internal pull-up resistors on the 20-pin devices. It is also used as the positive input of the analogue comparator (AIN1) on the 20-pin device.

P1.2 to P1.7

These are the remaining bi-directional I/O pins of port 1. These pins have internal pull-up resistors.

VCC

Supply voltage.

P0.0 to P0.7

These are the eight I/O pins of port 0 of the standard 8051. These pins have no pull-up resistors. P0.0 to P0.7 are also used to provide the low addresses (A0 to A7) and the data during fetches from external program memory and during accesses to external data memory.

P2.0 to P2.7

These are the eight I/O pins of port 2 of the standard 8051. These pins have pull-up resistors. P2.0 to P2.7 are also used to provide the high address (A8 to A15) byte during fetches from external program memory and during accesses to external data memory.

EA/VPP

This is the external access enable pin on the standard 8051. EA should be connected to VCC for internal program executions. This pin also receives the programming voltage during programming.

PSEN

This is the program store enable pin on the 8051 devices. This pin is activated when the device is executing code from external memory.

ALE/PROG

This is the address latch enable pin on the standard 8051 devices. This pin is used to latch the low byte of the address during accesses to external memory.

1.7 Timer/Counters

The 8051 and AT89C2051 contain two timer/counters known as timer/counter 0 and timer/counter 1 (larger members of the 8051 family contain more timers/counters). These timer/counters can be operated in several different modes depending upon the programming of two registers TCON and TMOD, as shown in Tables 1.2 and 1.3. These registers should be programmed before using any timer or counter facilities of the microcontroller.

Table 1.2 TCON timer/counter control register

Bit name	Bit position	Description
TF1	7	Timer 1 overflow flag. Set and cleared by hardware
TR1	6	Timer 1 run control bit. Timer 1 is turned on when TR1 = 1, and stopped when TR1 = 0
TF0	5	Timer 0 overflow flag. Set and cleared by hardware
TR0	4	Timer 0 run control bit. Timer 0 is turned on when TR0 = 1, and stopped when TR0 = 0
IE1	3	External interrupt 1 edge flag. Set and cleared by hardware
IT1	2	External interrupt 1 type. IT1 = 1 specifies interrupt on falling edge. IT1 = 0 specifies interrupt on low level
IE0	1	External interrupt 0 edge flag. Set and cleared by hardware
IT0	0	External interrupt 0 type. IT0 = 1 specifies interrupt on falling edge. IT0 = 0 specifies interrupt on low level

Table 1.3 TMOD timer/counter mode control register

TIMER 1				TIMER 0			
GATE	C/T	M1	M0	GATE	C/T	M1	M0

GATE: When TRx is set and GATE = 1, TIMER/COUNTERx runs only while the INTx pin is high. When GATE = 0, TIMER/COUNTERx will run only while TRx = 1.

C/T: Timer or counter select bit. When C/T = 0, operates as a timer (from internal clock). When C/T = 1, it operates as a counter (input from Tx input).

M1, M0:   Timer/counter mode select bits are defined in Table 1.4.

TCON is the timer/counter control register and this register is bit addressable.

Table 1.4 M1, M0 mode control bits

M1	M0	Operating mode
0	0	13-bit timer
0	1	16-bit timer/counter
1	0	8-bit auto-reload timer/counter
1	1	Two 8-bit timers

For example, bit 4 of TCON is the counter 0 run control bit and setting this bit starts counter 0. TCON register is at address 88 (hex) and bits in this register can be accessed either by making reference to the address or by using compiler reserved names (e.g. TR0).

TMOD is the timer/counter mode control register. This register sets the operating modes of the two timer/counters as shown in Table 1.3. There are three operating modes, known as modes 0, 1, and 2. TMOD is not bit addressable and should be loaded by specifying all the 8 bits. For example, loading hexadecimal byte 01 into TMOD sets timer 0 into mode 1 which is a 16-bit timer and is turned on and off by bit TR0 of TCON. Also, timer 1 is set into mode 0 which is a 13-bit timer and is turned on and off by bit TR1 of TCON.

1.8 Interrupt Control

The standard 8051 and AT89C2051 provide six interrupt sources:

Table 1.5 Interrupt entry locations in memory

Interrupt source	Interrupt number	Location in memory (hex)
External interrupt 0	0	0003
Timer 0	1	000B
External interrupt 1	2	0013
Timer 1	3	001B
Serial port	4	0023

Table 1.6 Interrupt enable/disable bits

EA	–	–	ES	ET1	EX1	ET0	EX0

Where:
EA: Global interrupt enable/disable. If EA = 0, no interrupt will be accepted. If EA = 1, each interrupt source is individually enabled or disabled by setting or clearing its bit, given below.
ES: Serial port interrupt enable bit.
ET1: Timer 1 interrupt enable bit.
EX1: External interrupt 1 enable bit.
ET0: Timer 0 interrupt enable bit.
EX0: External interrupt 0 enable bit.

- Two external interrupts (INT0 and INT1)
- Two timer interrupts (timer 0 and timer 1)
- One serial port receive interrupt
- One serial port transmit interrupt

Each interrupt is assigned a fixed location in memory and an interrupt causes the CPU to jump to that location, where it executes the interrupt service routine. Table 1.5 gives the interrupt sources and the start of their service routines in memory. Note that the serial port receive and transmit interrupts point to the same location.

Each interrupt source can be individually enabled or disabled by setting or clearing its interrupt enable bit. Table 1.6 gives the interrupt enable bit patterns.

1.9 Minimum Microcontroller Configuration

The minimum microcontroller configurations of the 8051- and AT89C2051-based microcontroller systems are shown in Figs 1.3 and 1.4. As can be seen from these figures, only the following external components are required to have a working microcontroller:

X1	Crystal (e.g. 12 MHz)
C1, C2	33 pF capacitors
C3	10 µF, 10 V electrolytic capacitor
R1	8.2K, 0.125 W resistor

We shall be using the circuit in Fig. 1.4 in all of the projects described in this book, except the last project which is based on a 40-pin device. The crystal chosen for the projects is 12 MHz, which gives a basic instruction timing of 1 µs. The power supply current of the AT89C2051 is around 15 mA, but a power supply which can deliver up to a few hundred milliamperes is recommended so that the interface circuitry attached to the microcontroller can be powered.

1.10 Project Development

Development of a AT89C2051 microcontroller project requires several development tools. The following is a list of the tools that are essential:

- Suitable assembler or compiler which can generate machine code for the AT89C2051 microcontroller. In this book we shall be developing the projects using a C compiler.
- Chip programmer suitable to program AT89C2051 devices. There are many programmers available on the market for this purpose. For example, PG302 by Inguana labs, Evalu8r by Equinox Technologies, and others. The programmer should be compatible with the code generated by the assembler or the compiler so that the code can be downloaded to the microcontroller. Notice that there is no ultraviolet erasing process. AT89C2051 devices contain reprogrammmable flash memories which can be erased and reprogrammed by electrical signals.
- A minimum AT89C2051 microcontroller hardware. Many manufacturers offer development systems, consisting of a basic microcontroller, LED lights, switches, buzzers etc. Some development systems include both language compilers and hardware and such systems can be very useful during project development.

Although the microcontroller used in the projects is the 20-pin AT89C2051, the code given will run on all members of the 8051 family provided that there is enough program and data memories.

CHAPTER 2

PROGRAMMING MICROCONTROLLERS IN C

The C programming language is a general-purpose high-level programming language that offers efficient and compact code and provides elements of structured programming. Many control and monitoring-based applications can be solved more efficiently with C than with any other programming language. C was originally available on mainframe computers, mini-computers, and personal computers (PCs). The C programming language is now available on most microcontrollers and microprocessors.

This book is not intended for teaching the C programming language. It is assumed that the reader is familiar with programming in C. The aim of this chapter is to show the special features of the C language when programming microcontrollers. In this book, the industry standard C51 optimizing C compiler is used throughout. This compiler has been developed by Keil Elektronik GmbH. C51 is available on both MS-DOS and Windows-based operating systems and the compiler implements the American National Standards Institute (ANSI) standard for the C language.

There are many other high-level language compilers available for microcontrollers, including **PASCAL**, **BASIC**, and other C compilers. Some of these compilers are freely available as shareware products and some can be obtained from the Internet with little cost. Also, some companies supply free limited capability compilers, mainly for evaluation purposes. These compilers can be used for learning the features of a specific product and in some cases small projects can be developed with such compilers. Section 2.5 gives a list of some sites where readers can find more information on high-level microcontroller compilers.

The C51 compiler has been developed for the 8051 family of microcontrollers. This is one of the most commonly used industry standard C compilers for the 8051 family, and can generate machine code for most of the 20-pin and 40-pin 8051 devices and its derivatives, including the following microcontrollers:

Intel and others 8051, 80C51, and 87C51
Atmel 89C51, 89C52, 89C55, 89S8252, and 89S53
Atmel 89C1051 and 89C2051

AMD 80C321, 80C521, and 80C541
Dallas 80C320, 80C520, and 80C530
Signetics 8xC750, 8xC751, and 8xC752
Siemens 80C517 and 80C537

C51 is a professional, industry standard compiler with many features, including a large number of built-in functions. In this chapter we shall be looking at the features of the C51 programming language as applied to programming single chip microcontrollers. More information on the C51 compiler is available from Keil Elektronik GmbH (see the *C51 Optimizing 8051 Compiler and Library Reference Manual*).

2.1 Data Types

The C51 compiler provides the standard C data types and in addition several extended data types are offered to support the 8051 microcontroller family. Table 2.1 lists the available data types (see the C51 reference manual for more information).

Some of the data types are described below in more detail.

2.1.1 bit

These data types may be used to declare 1-bit variables.

Example:

```
bit my_flag;        /* declare my_flag as a bit variable */
my_flag = 1;        /* set my_flag to 1 */
```

2.1.2 signed char/unsigned char

These data types are as in standard C language and are used to declare signed and unsigned character variables. Each character variable is 1 byte long (8 bits). Signed character variables range from -128 to $+127$; unsigned character variables range from 0 to 255.

Example:

```
unsigned char var1,var2;    /* declare var1 and var2 as unsigned char */
var1 = 0xA4;                /* assign hexadecimal A4 to variable var1 */
var2 = var1;                /* assign var1 to var2 */
```

Table 2.1 CSl data types

Data type	Bits	Range
bit	1	0 or 1
signed char	8	−128 to +127
unsigned char	8	0 to +255
enum	16	−32768 to +32767
signed short	16	−32786 to +32767
unsigned short	16	0 to 65535
signed int	16	−32768 to +32767
unsigned int	16	0 to 65535
signed long	32	−2147483648 to 2147483647
unsigned long	32	0 to 4294967295
float	32	±1.175494E-38 to ±3.402823E+38
sbit	1	0 or 1
sfr	8	0 to 255
sfr16	16	0 to 65535

2.1.3 signed short/unsigned short

These data types are as in standard C language and are used to declare signed and unsigned short variables. Each short variable is 2 bytes long (16 bits). Signed short variables range from −32 768 to +32 767 and unsigned short variables range from 0 to 65 535.

Example:

```
unsigned short temp;     /* declare temp as unsigned short */
unsigned short wind;     /* declare wind as unsigned short */
temp = 0x0C200;          /* assign hexadecimal C200 to variable temp */
wind = temp;             /* assign variable temp to wind */
```

2.1.4 signed int/unsigned int

As in the standard C language, these data types are used to declare signed and unsigned integer variables. Integer variables are 2 bytes long (16 bits). Signed integers range from −32 768 to +32 767 and unsigned integers range from 0 to 65 535.

Example:

```
unsigned int n1,n2;     /* declare n1 and n2 as unsigned integers */
n1 = 10;                /* assign 10 to n1 */
n2 = 2*n1;              /* multiply n1 by 2 and assign to n2 */
```

2.1.5 signed long/unsigned long

These data types are as in standard C language and are used to declare signed and unsigned long integer variables. Each long integer variable is 4 bytes long (32 bits).

Example:

```
unsigned long temp;     /* declare temp as long integer variable */
temp = 250000;          /* assign 250 000 to variable temp */
```

2.1.6 float

This data type is used to declare a floating point variable.

Example:

```
float t1,t2;            /* declare t1 and t2 as floating point variables */
t1 = 25.4;              /* assign 25.4 to t1 */
t2 = sqrt(t1);          /* assign the square-root of t1 to t2 */
```

2.1.7 sbit

This data type is provided for the 8051 family and is used to declare an individual bit within the SFR of the 8051 family. For example, using this data type one can access individual bits of an I/O port.

Example:

```
sbit switch = P1^3;     /* variable switch is assigned to bit 3 of port 1 */
switch = 0;             /* clear bit 3 of port 1 */
```

2.1.8 sfr

This data type is similar to sbit but is used to declare 8-bit variables.

Example:

```
sfr P1 = 0x90;                /* Port 1 address 0x90 assigned to P1 */
sfr P2 = 0xA0;                /* Port 2 address 0xA0 assigned to P2 */
unsigned char my_data;        /* declare my_data as unsigned character */
my_data = P1;                 /* read 8 bit data from port 1 and assign to my_data */
P2 = my_data++;               /* increment my_data and send to port 2 */
```

2.1.9 sfr16

This data type is similar to sfr but is used to declare 16-bit variables. When using this data type, the low byte should precede the high byte.

Example:

Timer 2 of the 8052 microcontroller uses addresses 0xCC and 0xCD for the low and high bytes. We can declare variable T2 to access both timer locations.

```
sfr16 T2 = 0xCC;              /* Timer 2, T2L=CC and T2H=CD */
T2 = 0xAE01;                  /* load Timer 2 with hexadecimal value AE01 */
```

2.2 Memory Models

8051 architecture supports both program (or code) and data memory areas. Program memory is read-only and it cannot be written to. Depending upon the type of processor used different amounts of internal program memory are available. For example, 8051 provides 4 Kbytes of internal program memory. Similarly, 89C2051 provides only 2 Kbytes of internal program memory. The program memory can be increased by connecting additional external memory to the basic microcontroller. There may be up to 64 Kbytes of program memory.

Data memory resides within the 8051 CPU and can be read from and written into. Up to 256 bytes of data memory are available depending upon the type of microcontroller used.

The memory model determines what type of program memory is to be used for a given application. There are three memory models, known as SMALL, COMPACT, and LARGE, and the required model is specified using the compiler directives. The SMALL memory model is used if all the variables reside in the internal data memory of the 8051. This memory model generates the fastest and the most efficient code and should be used whenever possible. In the COMPACT memory model, all variables reside in one page of external data memory. A maximum of 256 bytes of variables can be used. This memory model is not as efficient as the SMALL model. In the LARGE memory model, all variables reside in external data memory. A maximum of 64 Kbytes of data can be used. The LARGE model generates more code than the other two models and thus it is not very efficient.

Compiling in the SMALL memory model always generates the fastest and the smallest code possible since accessing the internal memory is always faster than accessing any external memory.

2.3 Interrupts

The C51 compiler allows us to declare interrupt service routines (ISRs) in our C code and then the program automatically jumps to this code when an interrupt occurs. The compiler automatically generates the interrupt vectors and entry and exit code for interrupt routines.

An ISR is declared similar to a function declaration but the interrupt number is specified as part of the function declaration. For example, the following is a declaration of the ISR for timer 1 interrupts (interrupt number 3):

```
Void timer1() interrupt 3
{
    interrupt service code goes in here
}
```

Similarly, the ISR for timer 0 (interrupt number 1) is declared as:

```
void timer0() interrupt 1
{
    interrupt service code goes in here
}
```

Note that we can specify the register bank to be used for the ISR with the **using** function attribute:

```
void timer0() interrupt 1 using 2
{
    interrupt service code goes in here
}
```

2.4 Structure of a Microcontroller-based C Program

The structure of a C program developed for a microcontroller is basically the same as the structure of a standard C program, with a few minor changes. The structure of a typical microcontroller-based C program is shown in Fig. 2.1. It is always advisable to describe the project at the beginning of a program using comment lines. The project name, filename, date, and the target processor type should also be included in this part of the program. The register definition file should then be included for the type of target processor used. This file is supplied as part of the compiler and includes the definitions for various registers of the microcontroller. In the example in Fig. 2.1, the register definition file for the Atmel 89C2051 type microcontroller is included. The global definitions of the variables used should then be entered, one line for each definition. The functions used in the program should then be included with the appropriate comments added to the heading and also to each line of the

```
/******************************************************************
Project:              Give project name
File:                 Give filename
Date:                 Date program was developed
Processor:            Give target processor type

This is the program header. Describe your program here briefly.
******************************************************************/
#include <AT892051.h>

#define ......           /* include your define statements here */

sbit ......             /* include your bit definitions here */

int ......
char ......             /* include your global declarations here */

void func1()            /* include you functions here */
{
}


main()                  /* main code */
{
                        /* include comments here */

}
```

Figure 2.1.
Structure of a microcontroller C program

functions. The main program starts with the keyword main(), followed by the opening brackets '{'. The lines of the main program should also contain comments to clarify the operation of the program. The program is terminated by a closing bracket '}'.

An example program is shown in Fig. 2.2. This program receives an 8-bit data from port 1 of an 89C2051 type microcontroller. The state of a switch, connected to bit 0 of port 3, is then checked. If the switch is 1, the value of the data is doubled by calling function *double_it*. If, on the other hand, the state of the switch is 0, the data value is incremented by 2 by calling function *inc_by2*, and then the program stops. It is important to realize that there is no returning point in a microcontroller program. Thus, where necessary, an endless loop should be formed at the end to stop the program from going into undefined parts of its code memory.

2.5 Program Description Language (PDL)

There are many methods that a programmer may choose to describe the algorithm to be implemented by a program. Flow charts have been used extensively in the past in many computer programming tasks. Although flow charts are useful, they tend to create an unstructured code and also a lot of time is usually wasted drawing them, especially when developing complex programs. In this section we shall be looking at a different way of describing the operation of a program, namely by using a program description language (PDL).

A PDL is an English-like language which can be used to describe the operation of a program. Although there are many variants of PDL, we shall be using simple constructs of PDL in our programming exercises, as described below.

2.5.1 START-END

Every PDL program (or sub-program) should start with a START statement and terminate with an END statement. The keywords in a PDL code should be highlighted in bold to make the code more clear. It is also good practice to indent program statements between the PDL keywords.

Example:

START
....
....
END

```
/*******************************************************************
Project:            A simple test
File:               TEST.C
Date:               10 August 1999
Processor:          89C2051

This program receives an 8-bit data from port 1 of the microcontroller and stores this data
in variable first. The state of a switch, connected to bit 0 of port 3, is then checked. If the
switch is 1, variable first is doubled by calling function double_it. If, on the other hand, the
state of the switch is 0, variable first is incremented by 2 by calling to inc_by2
*******************************************************************/
#include <AT892051.h>

#define ON 1
#define OFF 0

sbit switch = P3^0;             /* switch is connected to bit 0 of port 3 */

/* Function to double a value */
unsigned char double_it(unsigned char x)
{
   return (2*x);
}


/* Function to increment a value by 2 */
unsigned char inc_by2(unsigned char x)
{
   return (x+2);
}


/* Start of MAIN program */
main()
{
   unsigned char first,second;

   first = P1;                  /* get 8-bit data from port 1 */
   if(switch = = ON)
      second=double_it(first);  /* double the data if switch = 1 */
   else
      second=inc_by2(first);    /* otherwise increment by 2 */
   for(;;)                      /* wait here forever */
   {
   }
}                               /* end of MAIN program */
```

Figure 2.2.
Example of a microcontroller C program

2.5.2 Sequencing

For normal sequencing in a program, write the steps as short English text as if you are describing the program.

Example:

>   Turn on the valve
>   Clear the buffer
>   Turn on the LED

2.5.3 IF-THEN-ELSE-ENDIF

Use IF, THEN, ELSE, and ENDIF statements to describe the flow of control in your programs.

Example:

>   **IF** switch = 1 **THEN**
>   Turn on buzzer
>   **ELSE**
>   Turn off buzzer T
>   urn off LED
>   **ENDIF**

2.5.4 DO-ENDDO

Use DO and ENDDO control statements to show iteration in your PDL code.

Example:

>   Turn on LED
>   **DO** 5 times
>   Set clock to 1
>   Set clock to 0
>   **ENDDO**

A variation of the DO-ENDDO construct is to use other keywords like DO-FOREVER, DO-UNTIL etc. as shown in the following examples.

>   Turn off the buzzer
>   **IF** switch = 1 **THEN**
>   **DO UNTIL** Port 1 = 2
>   Turn on LED
>   Read port 1
>   **ENDDO**
>   **ENDIF**

or

> **DO FOREVER**
>   Read data from port 1
>   Display data
>   Delay a second
> **ENDDO**

2.5.5 REPEAT-UNTIL

This is another useful control construct which can be used in PDL codes. An example is shown below where the program waits until a switch value is equal to 1.

> **REPEAT**
>   Turn on buzzer
>   Read switch value
> **UNTIL** switch = 1

2.6 Internet Web Sites of Microcontroller Compilers

The amount of microcontroller software available on the Internet is huge and there are many different example programs. Internet web sites of some popular 8051 family microcontroller compilers and other useful sites are given below.

Pascal compilers

Embedded Pascal – 8051/8051
http://www.grifo.it/SOFT/Lawicel/uk_EP_51.htm

Pascal 51
http://www.grifo.com/SOFT/KSC/Pascal51.htm

Embedded Pascal
http://www/users.iafrica.com/r/ra/rainier/index.htm

ElProg Pascal51
http://www.geocoties.com/SiliconValley/Campus/9592/index.html

SYSTEM51 Pascal
http://www.spacetools.de/tools/space-program/space/products/s_050006.htm

C Compilers

MICRO/C-51
htp://www.mcc-us.com/51tools.htm

Small C
http://www.newmicros.com/smallc51.html

IDE51-C
http://www.spjsystems.com/ide51.htm

SDCC (freeware 8051 C compiler)
http://www.geocoties.com/ResearchTriangle/Forum/1353/

C51
ttp://www.keilsoftware.com/home.htm

Various BASIC and C compilers
http://www.equinox-tech.com

MICRO-C
http://www.dunfield.com/dks.htm

Basic compilers

BASCOM
http://www.x54all.nl/~mcselec/bascom.html

TINY BASIC
http://www.code.archive.aisnota.com/

BASIKIT
http://www.mdllabs.com/basikit.htm

BXC-51
http://www.mindspring.com/~tavve/8051/bxc51.html

BEC-51
http://www.windspring.com/~tavve/8051/bec51.html

Useful site on 8051 software and hardware
http://www.cis.ohio-state.edu/hypertext/faq/usenet/microcontroller-faq/8051/faq.html

2.7 Further Reading

The following books and reference manuals are useful in learning to program in C.

The C Programming Language (2nd edn)
Kernighan & Richie
Prentice-Hall, Inc.
ISBN 0-13-110370-9

C and the 8051: Programming and Multitasking
Schultz
PTR Prentice-Hall, Inc.
ISBN 0-13-753815-4

C for Dummies
Dan Gookin
ISBN 1-878058-78-9

C The Complete Reference
Herbert Schildt
ISBN 0-07-882101-0

Efficient C
Plum & Brodie
Plum Hall Inc.
ISBN 0-911537-05-8

C51 Compiler, Optimizing 8051 C Compiler and Library Reference
User's Guide
Keil Elektronik GmbH

CHAPTER 3

LIGHT PROJECTS

This chapter describes simple light projects using the basic 89C2051 microcontroller circuit described in earlier chapters. Over ten projects are given, from very simple LED display projects to complex projects incorporating alphanumeric displays. For each project, the following information is given as appropriate:

- *Function*: what the project does, its inputs and outputs.
- *Circuit diagram*: full circuit diagram of the project and explanation of how the circuit works.
- *Program description*: functional description of the software in simple English-like language (PDL).
- *Program listing*: full tested and working C program listing for each project, including comments.
- *Components required*: listing of components required to build each project.

PROJECT 1 – LED Binary Counter

Function

This project counts up in binary and displays the result on eight LEDs connected to port 1 of the microcontroller as shown in Fig. 3.1.

Circuit Diagram

As shown in Fig. 3.2 the circuit is extremely simple, consisting of the basic 89C2051-based microcontroller and eight LEDs connected to port 1 of the microcontroller. Each microcontroller output pin can sink a maximum of 80 µA and source up to 20 mA. The manufacturers specify that the total source current of a port should not exceed 80 mA. There are many different types of LED lights on the market, emitting red, green, amber, white, or yellow colours. Standard red LEDs require about 5 to 10 mA to emit visible bright light. There are also low-current small LEDs operating from as low as 1 mA.

30 Microcontroller Projects in C for the 8051

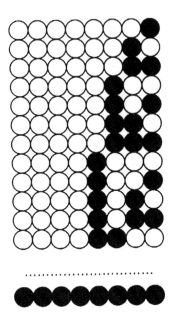

Figure 3.1.

Output pattern of Project 1

In Fig. 3.2, the microcontroller outputs operate in current source mode where an LED is turned on if the corresponding output is at logic LOW level. The required value of the current limiting resistors can be calculated as follows:

$$R = \frac{V_s - V_f}{I_f}$$

where V_s is the supply voltage (+5 V), V_f is the LED forward voltage drop (about 2 V), and I_f is the LED forward current (1 to 30 mA depending on the type of LED used). In this design if we assume an LED current of about 6 mA, the required resistors will be:

$$R = \frac{5-2}{6} \cong 470 \, \Omega$$

Although eight individual resistors are shown in this circuit, it is possible to replace these resistors with a single DIL (dual-in-line) resistor chain.

Program Description

The program is required to increment a value and then output to port 1 of the microcontroller. Because the microcontroller operates at a very high speed, it is

Light Projects

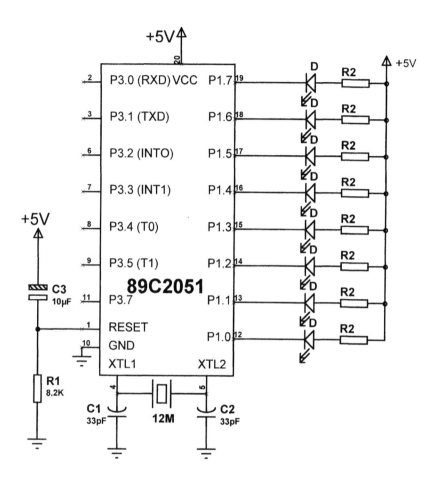

Figure 3.2.
Circuit diagram of Project 1

necessary to insert a delay in the program so that the LED outputs can be seen visually. The following PDL describes the functions of the program:

START
    Set count to 1
    DO FOREVER
        Output count to port 1
        Increment count
        Delay
    ENDDO
END

```
/****************************************************************
PROJECT:              PROJECT 1
FILE:                 PROJ1.C
DATE:                 August 1999
PROCESSOR:            AT892051

This project counts up in binary and displays the result on eight LEDs connected to port 1.
The data is displayed with about 1 second delay between each output.
****************************************************************/
#include <AT892051.h>

/* Function to delay about a second */
void wait_a_second()
{
   unsigned int x;
   for(x=0;x<33000;x++);
}

/* Start of main program */
main()
{
   int LED=1;                    /*initialize count to 1*/

   for(;;)                       /*Start of endless loop*/
   {
      P1=~LED;                   /*Invert and output*/
      LED++;                     /*Increment the count*/
      wait_a_second();           /*Wait about a second*/
   }
}
```

Figure 3.3.

Program listing of Project 1

Program Listing

The full program listing is shown in Fig. 3.3. Variable LED is initialized to 1 and is used as the counter. The endless loop is set using the *for* statement with no parameters. Variable *P1* is defined in include file 'AT892051.h' and this is a reserved name for port 1 of the microcontroller. Notice that variable LED is complemented (using operator '~') and then sent to the output port. This is necessary since the output ports are configured to source current, i.e. an LED is turned on when the corresponding port output is logic LOW. A delay of approximately 1 second is obtained by the function *wait_a_second*. This function is simply a dummy *for* loop and gives about 1 second delay when the microcontroller is operated with a 12 MHz crystal. Different values of loop

Light Projects

count will give different delays. Also, different delays will be obtained with other C compilers. More accurate and compiler independent delays can be obtained using the timer utilities of the microcontroller.

Components Required

In addition to the components required by the basic microcontroller circuit, the following components will be required for this project:

R2   470 Ω, 0.125 W resistor (8 off), or DIL package
D   LED (8 off)

PROJECT 2 – LED Chasing Circuit

Function

This project turns on the LEDs connected to port 1 of the microcontroller in sequence, resulting in a chasing LED effect. The data is displayed with about 1 second delay between each output pattern. Figure 3.4 shows the output pattern displayed by the LEDs.

Circuit Diagram

The same circuit (Fig. 3.2) as in Project 1 is used. The LEDs can be mounted in a circular or in some other geometric form to enhance the chasing effect. Also, different coloured LEDs can be used to give a colourful output.

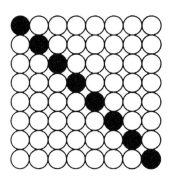

Figure 3.4.
Output pattern of Project 2

Program Description

The program is required to load a 1 into the top (or bottom) bit of a variable and then shift the data right (or left) by one digit and display on the LEDs. A delay will be required between each output. The following PDL describes the functions of the program. In this PDL, the top bit of a variable is loaded and data is shifted to the right:

START
    Set count to 128
    DO FOREVER
        Output count to port 1
        Shift count right by 1 digit
        IF count = 0 **THEN**
            Set count back to 128
        ENDIF
        Delay
    ENDDO
END

Program Listing

The full program listing is shown in Fig. 3.5. Variable *LED* is initialized to 128 (top bit set to 1) and is used as the counter. The endless loop is set using the *for* statement with no parameters. Variable *P1* is defined in include file 'AT892051.h' and this is a reserved name for port 1 of the microcontroller. Notice that variable LED is complemented and then sent to the output port. This is necessary since the output ports are configured to source current, i.e. an LED is turned on when the corresponding port output is logic LOW. The count value is shifted right by one digit using the C compiler operator '≫'. When the count reaches 0, it is reloaded with 128, i.e. the values of variable *LED* will be 128, 64, 32, 16, 8, 4, 2, 1, 128, ... A delay of approximately 1 second is obtained by the function *wait_a_second*.

PROJECT 3 – Random LED Pattern

Function

This project turns on the LEDs connected to port 1 randomly. A random number is generated between 0 and 32 767 using the built-in C function *rand* and then this is output to turn on the corresponding LEDs. The data is displayed with about 1 second delay between each output pattern.

```
/***********************************************************************
PROJECT:              PROJECT 2
FILE:                 PROJ2.C
DATE:                 August 1999
PROCESSOR:            AT892051

This project turns on the LEDs connected to port 1 in sequence, resulting in a chasing LED
effect. The data is displayed with about 1 second delay between each output.
***********************************************************************/

#include <AT892051.h>

/* Function to delay about a second */
void wait_a_second()
{
  unsigned int x;
  for(x=0;x<33000;x++);
}

/* Start of main program */
main()
{
  unsigned char LED=128;      /*initialize to 128*/
  for(;;)                     /*Start of loop*/
  {
    P1=~LED;                  /*Invert and output*/
    LED=LED >> 1;             /*Shift to right*/
    if(LED == 0)LED=128;      /*Set to 128*/
    wait_a_second();          /*Wait a second*/
  }
}
```

Figure 3.5.

Program listing of Project 2

Circuit Diagram

The same circuit (Fig. 3.2) as in Project 1 is used. The LEDs can be mounted in different patterns and in different colours depending upon the application.

Program Description

The program is required to generate a random number and then output this number to port 1 in order to turn on the corresponding LEDs. A small delay is required between each output so that the LED patterns can be seen. The following PDL describes the functions of the program:

```
START
    DO FOREVER
        Generate a random number
        Output number to port 1
        Delay
    ENDDO
END
```

Program Listing

The full program listing is shown in Fig. 3.6. Variable *LED* is used to hold the data. The endless loop is set using the *for* statement with no parameters.

```
/******************************************************************
PROJECT:            PROJECT 3
FILE:               PROJ3.C
DATE:               August 1999
PROCESSOR:          AT892051

This project turns on the LEDs connected to port 1 randomly. A random number is
generated between 0 and 32767 and then this is output to turn on the corresponding
LEDs. The data is displayed with about 1 second delay between each output.
*******************************************************************/

#include <stdlib.h>
#include <AT892051.h>

/* Function to delay about a second */
void wait_a_second()
{
    unsigned int x;
    for(x=0;x<33000;x++);
}

/* Start of main program */
main()
{
    int LED;

    for(;;)                     /*Start of endless loop*/
    {
        /* Generate a random number between 0 and 32767 */
        LED=rand();
        P1=~LED;                /*Invert and output*/
        wait_a_second();        /*Wait a second*/
    }
}
```

Figure 3.6.
Program listing of Project 3

Variable *P1* is defined in include file 'AT892051.h' and this is a reserved name for port 1 of the microcontroller. The built-in C function *rand()* generates a random integer number between 1 and 32 767 and this function is used to generate a random number and store it in variable *LED*. The generated number is complemented and output to port 1 of the microcontroller and the process is repeated indefinitely with about 1 second delay between each output pattern.

PROJECT 4 – Cyclic LED Pattern

Function

This project turns on the LEDs connected to port 1 in a cyclic manner such that first only 1 LED is on, then 2 LEDs are on, then 3, 4, 5, ... , 8 are on (Fig. 3.7). The process is repeated indefinitely with 1 second delay between each output pattern.

Circuit Diagram

The same circuit (Fig. 3.2) as in Project 1 is used. The LEDs can be mounted in different patterns and in different colours depending upon the application.

Program Description

The program is required to turn on the first LED (e.g. corresponding to number 128) and then after a second delay turn on the LEDs corresponding to

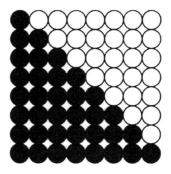

Figure 3.7.

Output pattern of Project 4

numbers 64, 32, 16 and so on until all eight LEDs are on (number 255). The process is then repeated forever as shown in Fig. 3.7 with about 1 second delay between each output pattern. The following PDL describes the functions of the program:

START
    Set count to 128
    DO FOREVER
        Output count to port 1
        Shift count to right by 1 digit
        IF count = 255 **THEN**
            Delay
            Output to port 1
            Set count to 128
        ENDIF
        Delay
    ENDDO
END

Program Listing

The full program listing is shown in Fig. 3.8. Variable *LED* is initialized to 128 (top bit on) and used to hold the data. This value is complemented and output to port 1, and then shifted right by 1 digit using the C operator '≫'. When all the LEDs are on (LED = 255), the last value in the chain is displayed and variable *LED* is set back to 128. The above process is repeated forever with about 1 second delay between each output pattern.

PROJECT 5 – LED Dice

Function

This project simulates a dice by displaying a random number between 1 and 6, on six LEDs connected to port 1 of the microcontroller. Bit 0 of port 3 (P3.0) is used as the input and a push-button switch is connected to this pin. Every time the switch is pressed, a new number is displayed.

Circuit Diagram

The circuit diagram of this project is shown in Fig. 3.9. Bit 0 of port 3 is normally held at logic HIGH with the pull-up resistor R3. When switch S1 is pressed, bit 0 of port 3 moves to logic LOW and is detected by the software. As

```
/****************************************************************************
PROJECT:                PROJECT 4
FILE:                   PROJ4.C
DATE:                   August 1999
PROCESSOR:              AT892051

This project turns on the LEDs connected to port 1 in a cyclic manner such that first only 1
LED is on, then 2, 3, 4, 5, ... , 8 are on. The process is repeated. The data is displayed with
about 1 second delay between each output.
****************************************************************************/

#include <AT892051.h>

/* Function to delay about a second */
void wait_a_second()
{
  unsigned int x;
  for(x=0;x<33000;x++);
}

/* Start of main program */
main()
{
  unsigned char LED=128;      /*Initialize count*/

  for(;;)                     /*Start of loop*/
  {
    P1=~LED;                  /*Invert and output*/
    LED=LED | (LED >> 1);     /*Obtain next value*/
    if(LED == 255)            /*If end of pattern ... */
    {
      wait_a_second();
      P1=~LED;
      LED=128;
    }
    wait_a_second();          /*Wait a second*/
  }
}
```

Figure 3.8.

Program listing of Project 4

shown in Fig. 3.9, the seven LEDs have been mounted in a pattern to emulate the dots on a real dice. The pattern displayed for different numbers is shown in Fig. 3.10. As in a real dice, the first row can have up to two LEDs on (corresponding to two dots on a dice), the second row up to three LEDs on, and the third row can have up to two LEDs on.

40 Microcontroller Projects in C for the 8051

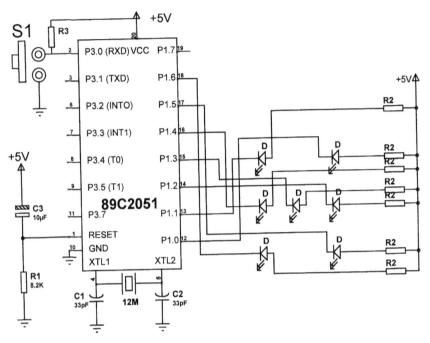

Figure 3.9.
Circuit diagram of Project 5

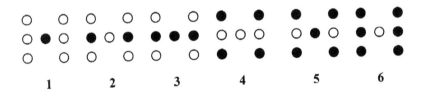

Figure 3.10.
LED pattern displayed for different dice numbers

Program Description

A random dice number is obtained during scanning of the push-button switch as follows. The program scans the push-button switch continuously. If the switch is not pressed (i.e. at logic HIGH), a number is incremented between 1 and 6. Whenever the push-button is pressed, the current value of the number is read and this value is used as the new dice number. Since the switch is pressed by the user in random, the numbers generated are also random numbers from 1 to 6. The new random number is displayed on the seven LEDs appropriately.

After about 2 seconds delay, all LEDs are turned off and the above process is repeated forever. The following PDL describes the functions of the program:

START
    Initialize count to 0
    DO FOREVER
        IF Push-button is pressed **THEN**
            Read the new dice number from count
            Turn on the appropriate dice LEDs
            Delay about 2 seconds
            Turn off all LEDs
        ELSE
            Increment count
            IF count = 7 **THEN**
                Count = 1
            ENDIF
        ENDIF
    ENDDO
END

Table 3.1 shows the random numbers generated and the corresponding LEDs that will be turned on to give the dice display of Fig. 3.10.

As an example, if the number 3 is to be displayed then only LEDs D3, D4, and D5 will be turned on. Similarly, for number 6, all LEDs except LED 4 will be turned on.

Program Listing

The full program listing is shown in Fig. 3.11. Variable *ON* is defined as 0 and

Table 3.1 Dice Numbers and corresponding LED patterns

Number	LED on
1	D4
2	D3, D5
3	D3, D4, D5
4	D1, D2, D6, D7
5	D1, D2, D4, D6, D7
6	D1, D2, D3, D5, D6, D7

```
/******************************************************************
PROJECT:            PROJECT 5
FILE:               PROJ5.C
DATE:               August 1998
PROCESSOR:          AT892051

This is a dice simulator project. Seven LEDs are connected to port 1 of the microcontroller
and arranged as shown in the text. A push-button switch is connected to bit 0 of port 3
and when this switch is pressed, a new number is obtained between 1 and 6 and the
corresponding LEDs are turned on to simulate a real dice. After 2 seconds delay, all LEDs
are turned off and the user can throw a dice again.
******************************************************************/
#include <AT892051.h>

#define ON 0
#define ALL_OFF 0xFF

sbit button = P3^0;           /*bit P3.0 is the push-button*/
sbit D1=P1^0;                 /*define dice patterns*/
sbit D2=P1^1;
sbit D3=P1^2;
sbit D4=P1^3;
sbit D5=P1^4;
sbit D6=P1^5;
sbit D7=P1^6;

/* Function to delay about a second */
void wait_a_second()
{
    unsigned int x;
    for(x=0;x<33000;x++);
}

/* Start of main program */
main()
{
    int DICE=0;               /*Initialize to 0*/

    for(;;)                   /*Start of loop*/
    {
        if(button == 0)       /*Button pressed*/
        {
            switch(DICE)
            {
                case 1:       /*DICE=1*/
                    D4=ON;
                    break;
```

```
        case 2:                 /*DICE=2*/
            D3=ON;
            D5=ON;
            break;
        case 3:
            D3=ON;               /*DICE=3*/
            D4=ON;
            D5=ON;
            break;
        case 4:
            D1=ON;               /*DICE=4*/
            D2=ON;
            D6=ON;
            D7=ON;
            break;
        case 5:                  /*DICE=5*/
            D1=ON;
            D2=ON;
            D4=ON;
            D6=ON;
            D7=ON;
            break;
        case 6:                  /*DICE=6*/
            D1=ON;
            D2=ON;
            D3=ON;
            D5=ON;
            D6=ON;
            D7=ON;
            break;
        }
        wait_a_second();         /*Wait 2 sec*/
        wait_a_second();
        P1=ALL_OFF;              /*LEDs OFF*/
    }
    else
    {
        DICE++;
        if(DICE = = 7)DICE=1;    /*Set to 0 if 7*/
    }
   }
 }
```

Figure 3.11.
Program listing of Project 5

variable *ALL_OFF* is defined as hexadecimal 0xFF (i.e. all bits set). These variables will be used to turn an LED on or to turn all LEDs off. Variable *button* is assigned to bit 0 of port 3 using the C compiler *sbit* statement. Similarly, the seven LEDs are assigned to bits 0 to 7 of port 1 using the *sbit* statement.

Variable *DICE* is declared as an integer and holds the dice values. Inside the endless *for* loop, variable *button* is tested. If *button* is 0, i.e. if the user presses the push-button, then a *switch* statement is used to turn on the appropriate LEDs as defined in Table 3.1. As an example, if the value of *DICE* is 2, LEDs D3 and D5 are turned on and the others are turned off. The *break* instructions ensure that we jump out of the *switch* statement after executing a case block. A dice value is displayed for 2 seconds and after this time all the LEDs are turned off. If inside the endless *for* loop the *button* is not pressed (i.e. *button* is 1), then variable *DICE* is incremented continuously. When *DICE* is 7, it is set back to 1.

A More Efficient Program

Notice that in this program we have used a *switch* statement and executed the correct case block depending on the value of variable *DICE*. We can make the program much more efficient and easy to follow if we create a simple table (an array in the program) and in this table store the dice numbers against the hexadecimal values of LED patterns. Table 3.2 shows the relationship between the dice numbers, the LED patterns and the corresponding binary and hexadecimal equivalents.

As an example, to display number 3 pattern, all we have to do is send hexadecimal number 1C to port 1. Similarly, sending 77 will display the dice pattern for number 6 on the LEDs.

The program listing given in Fig. 3.12 uses Table 3.2 to display dice patterns. The hardware setup is again the same and bit 0 of port 3 is used as the push-

Table 3.2 Dice numbers and corresponding bit patterns			
Number	LED on	Binary	Hex
1	D4	00001000	08
2	D3, D5	00010100	14
3	D3, D4, D5	00011100	1C
4	D1, D2, D6, D7	01100011	63
5	D1, D2, D4, D6, D7	01101011	6B
6	D1, D2, D3, D5, D6, D7	01110111	77

```
/****************************************************************
PROJECT:                PROJECT 5
FILE:                   PROJ5-1.C
DATE:                   August 1999
PROCESSOR:              AT892051

This is a dice simulator project. Seven LEDs are connected to port 1 of the microcontroller
and mounted as shown in the text. A push-button switch is connected to bit 0 of port 3
and when this switch is depressed, a new number is obtained between 1 and 6 and the
corresponding LEDs are turned on to simulate a dice. After 2 seconds delay, all LEDs are
turned off and the user can throw a dice again.

This code is more efficient than the previous dice code.
****************************************************************/
#include <AT892051.h>

#define ALL_OFF   0xFF

sbit button = P3^0;             /*Bit P3.0 is the push-button*/

/* Function to delay about a second */
void wait_a_second()
{
   unsigned int x;
   for(x=0;x<33000;x++);
}

/* Start of main program */
main()
{
   int DICE=0;                  /*Initialize to 0*/
   int DICE_ARRAY[6]={0x08,0x14,0x1C,0x63,0x6B,0x77};

   for(;;)                      /*Start of loop*/
   {
      if(button == 0)           /*Button pressed?*/
      {
         P1=~DICE_ARRAY[DICE-1];
         wait_a_second();       /*Wait 2 secs*/
         wait_a_second();
         P1=ALL_OFF;            /*turn off LEDs*/
      }
      else
      {
         DICE++;                /*Inc DICE*/
         if(DICE == 7)DICE=1;   /*Set to 0 if 7*/
      }
   }
}
```

Figure 3.12.
More efficient code for Project 5

button input. An integer array *DICE_ARRAY* is created to store the hexadecimal bit patterns as described in the table. Index 0 of this array corresponds to dice number 1 (pattern generated by hexadecimal number 8) and index 1 corresponds to dice number 2 (pattern generated by hexadecimal number 14) and so on. Because the arrays in C are indexed from 0, we have to subtract 1 from the array index in order to get the correct value. As shown in the program listing in Fig. 3.12, variable *button* is tested inside the endless loop. If the push-button is pressed, the hexadecimal bit pattern corresponding to variable *DICE-1* is obtained using the statement *DICE_ARRAY[DICE-1]* and this value is complemented and sent to port 1 of the microcontroller. All the displays are turned off after about 2 seconds delay. If the push-button is not pressed, variable *DICE* is incremented continuously and set back to 1 when it reaches 7.

Components Required

In addition to the components required by the basic microcontroller circuit, the following components will be required for this project:

R2   470 Ω, 0.125 W resistors (7 off)
R3   100K, 0.125 W resistor
D   LEDs (7 off)

PROJECT 6 – Hexadecimal Display

Function

This project shows how a microcontroller can be interfaced to a TIL311 type hexadecimal display. The program counts up from 0 to 9 and then in hexadecimal format from A to F and then back to 0. This process is repeated forever with a 1 second delay inserted between each count.

Circuit Diagram

The circuit diagram of this project is shown in Fig. 3.13. TIL311 is a popular 14-pin DIL small hexadecimal display, powered from +5 V. Inputs A, B, C, D of the display are the data inputs and these are connected to the lower part of port 1 (P1.0 to P1.3). LATCH input (pin 5) controls the display. When LATCH is LOW, new data is written to the display. When LATCH is high, the display data is frozen. The LATCH input of the display is connected to bit 7 of port 1 (P1.7). A new data is displayed by sending the data to the A, B, C, D

Light Projects 47

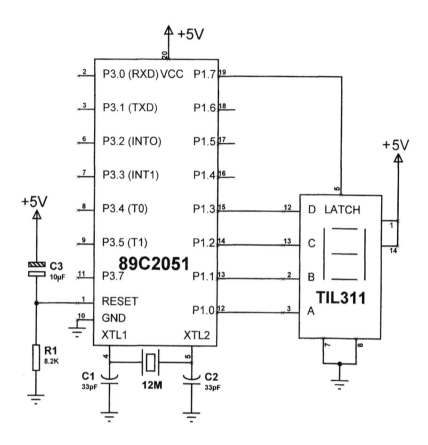

Figure 3.13.

Circuit diagram of Project 6

inputs and then the LATCH input is set to logic LOW and then back to HIGH. Pins 1 and 14 of the display are connected to +5V and pins 7 and 8 are connected to the ground.

Program Description

The program is very simple. The count is initially set to 0 and the display latch is set to 1 to avoid any unintentional write to the display. The count is then sent to the display and the display latch is clocked. The next data value is obtained by incrementing the count. When the count reaches 16, it is reset back to 0. The following PDL describes the functions of the program. Function *out_til311* displays data on the TIL311:

Main program

START
    Set count to 0
    Set display latch to 1
    DO FOREVER
        Call function out_til311 to display the count
        Increment the count
        IF count = 16 **THEN**
            Count = 0
        ENDIF
        Delay a second
    ENDDO
END

Function out_til311

Input: Count
Output: None

START
    Set top bit of count
    Send count to port 1
    Set LATCH to LOW
    Set LATCH to HIGH
END

Program Listing

The full program listing is shown in Fig. 3.14. Variable *latch* is assigned to bit 7 of port 1 using the *sbit* instruction of the compiler. Variable *CNT* is initialized to 0 and function *out_til311* is called to display the value of *CNT*. *CNT* is then incremented by 1. When *CNT* is 16, it is reset back to 0. The loop is repeated forever after a 1 second delay between each count. The displayed data is:

0 1 2 3 4 5 6 7 8 9 10 A B C D E F 0 1 ...

Function *out_til311* receives the data to be displayed as its argument (i.e. *x*). The *latch* is initially set to 1 by logical ORing the data with hexadecimal value 0x80. The *latch* is then set to 0 to enable the data to be written to the display and then back to 1 to freeze the display.

Light Projects   49

```
/******************************************************************************
PROJECT:                PROJECT 6
FILE:                   PROJ6.C
DATE:                   August 1999
PROCESSOR:              AT892051

This is a counter project. A TIL311 type hexadecimal alphanumeric display is connected
to port 1 of the microcontroller. The program counts from 0 to 9 and then from A to F
(hexadecimal). The data is displayed with about 1 second delay between each output.
******************************************************************************/
#include <AT892051.h>

sbit latch = P1^7;              /*Bit P1.7 is the latch*/

/* Function to delay about a second */
void wait_a_second()
{
  unsigned int x;
  for(x=0;x<33000;x++);
}

/* Function to display data on a TIL311 display */
void out_til311(int x)
{
  P1=x | 0x80;                  /*Send data with latch=1*/
  latch=0;                      /*Latch the data*/
  latch=1;                      /*Set latch on*/
}

/* Start of main program */
main()
{
  int CNT=0;                    /*Initialize count*/

  latch=1;                      /*Set latch on*/
  for(;;)                       /*Start of loop*/
  {
    out_til311(CNT);            /*Output to TIL311*/
    CNT++;                      /*Increment count*/
    if(CNT == 16)CNT=0;         /*Back to 0 if 16*/
    wait_a_second();            /*Wait a second*/
  }
}
```

Figure 3.14.

Program listing of Project 6

It is interesting to note that, in many programming applications a variable is incremented and then tested to see whether it reached a constant value. An example is given in Fig. 3.14 where variable CNT is used:

CNT++;
If(CNT == 16)CNT=0;

Note that the above C code could also be written in a more compact form as:

If(++CNT == 16)CNT=0;

Components Required

In addition to the components used for the basic microcontroller circuit, a TIL311 type hexadecimal display will be required for this project.

PROJECT 7 – Two-Digit Decimal Count

Function

This project shows how a microcontroller can be interfaced to two TIL311 type hexadecimal displays. This project counts up continuously from 0 to 99 in decimal with about a second delay between each count.

Circuit Diagram

The circuit diagram of this project is shown in Fig. 3.15. Two TIL311 type hexadecimal displays are used. Display MSD (most significant digit) will be programmed to show the tens and LSD (least significant digit) will show the units. Data inputs (A, B, C, D) of both displays are connected in parallel to the lower part of port 1 (P1.0 to P1.3). LATCH inputs (pin 5) of the displays are controlled separately. LATCH input of display MSD is connected to P1.7 and the same input of display LSD is connected to P1.6 of the microcontroller.

MSD data is displayed by sending the data to port 1 and then clocking pin P1.7. Similarly, LSD data is displayed by sending data to port 1 but this time clocking pin P1.6.

Light Projects 51

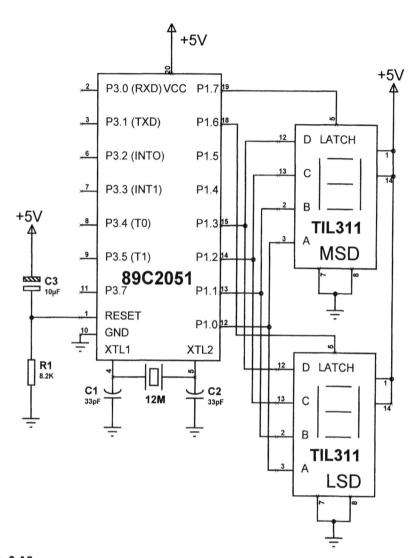

Figure 3.15.

Circuit diagram of Project 7

Program Description

The count is initially set to 0. The count is then sent to the display using a function called out2_til311. This function separates the variable into two decimal parts (MSD and LSD) and sends each part to the appropriate display. The next data value is obtained by incrementing the count. When the count reaches 100, it is reset back to 0. The following PDL describes the functions of the program.

Main program

START
    Set count to 0
    Set latches to 1
    DO FOREVER
        Call function out2_til311 with count to display the data
        Increment the count
        IF count = 100 **THEN**
            Count = 0
        ENDIF
        Delay a second
    ENDDO
END

Function out2_til311

Input: Count
Output: None

START
    Extract the first digit (MSD) of count
    Extract the second digit (LSD) of count
    Set top two bits of MSD
    Send MSD to port 1
    Set MSD LATCH to LOW
    Set MSD LATCH to HIGH
    Set top two bits of LSD
    Send LSD to port 1
    Set LSD LATCH to LOW
    Set LSD LATCH to HIGH
END

Program Listing

The full program listing is shown in Fig. 3.16. LATCH input of display MSD is named *latch_msd* and is assigned to port pin P1.7 using the *sbit* instruction. Similarly, LATCH input of display LSD is named *latch_lsd* and is assigned to port pin P1.6. The count (*CNT*) is initially set to 0 and both latches are set to 1 to avoid any accidental write to the displays. An endless loop is then formed using the *for* statement with no arguments. Function *out2_til311* is called inside the loop to display the value of *CNT*. *CNT* is then incremented by 1 and when it reaches 100, it is reset back to 0. The loop is repeated after a 1 second delay between each output value. The displayed data is:

0 1 2 3 4 5 6 7 8 9 10 ... 98 99 0 1 2 ...

Function *out2_til311* receives the data to be displayed (x) as its argument. This data is then divided by 10 and assigned to integer variable *msd* and is the data for the MSD display. The LSD data is calculated by subtracting $10*msd$ from input variable x and then assigning this to an integer variable named *lsd*. The top 2 bits of *msd* data are set to 1 by logical ORing the *msd* data with hexadecimal constant 0xC0. This freezes both displays and avoids any unwanted changes in the displayed data. The value of *msd* is then sent to port 1 of the microcontroller by clocking the *latch_msd*. Similarly, the top 2 bits of the *lsd* data are set to 1 to avoid any accidental write to the wrong display and then *lsd* is sent to port 1 of the microcontroller by clocking the *latch_lsd* bit.

Components Required

In addition to the components used for the basic microcontroller circuit, two TIL311 type hexadecimal displays will be required for this project.

PROJECT 8 – TIL311 Dice

Function

This project is a dice made up from a TIL311 type hexadecimal display. When a push-button switch, connected to bit 0 of port 3, is depressed, a random number between 1 and 6 is displayed on the display. After about 2 seconds the display is cleared and the user can throw a dice again. The program runs in an endless loop.

Circuit Diagram

The circuit diagram of this project is shown in Fig. 3.17. A TIL311 display is connected as in Project 6. Additionally, a push-button switch (S1) is connected to bit 0 of port 3. This pin is normally held at logic 1 with the pull-up resistor R2 and goes to logic 0 when the switch is pressed.

Program Description

The display latch is initially set to logic 1 to avoid any accidental data display. The state of push-button S1 is then checked continuously and when

/***
PROJECT: PROJECT 7
FILE: PROJ7.C
DATE: August 1999
PROCESSOR: AT892051

This is a dual display counter project. Two TIL311 type hexadecimal alphanumeric displays are connected to port 1 of the microcontroller.

The program counts from 0 to 99 and then back to 0. The data is displayed with about 1 second delay between each output.
***/
#include <AT892051.h>

sbit latch_msd = P1^7;   /*Bit P1.7 is the msd latch*/
sbit latch_lsd = P1^6;   /*Bit P1.6 is the lsd latch*/

/* Function to delay about a second */
void wait_a_second()
{
    unsigned int x;
    for(x=0;x<33000;x++);
}

/* Function to display data on two TIL311 displays */
void out2_til311(int x)
{
    int lsd,msd;
    msd=x/10;   /*Find msd*/
    lsd=x-10*msd;   /*Find lsd*/
    P1=msd | 0xc0;   /*Send msd data*/
    latch_msd=0;   /*Latch msd data*/
    latch_msd=1;   /*Set msd latch on*/
    P1=lsd | 0xc0;   /*Send lsd data*/
    latch_lsd=0;   /*Latch lsd data*/
    latch_lsd=1;   /*Set lsd latch on*/
}

/* Start of main program */
main()
{
    int CNT=0;   /*Initialize count*/

    latch_msd=1;   /*Set msd latch on*/
    latch_lsd=1;   /*Set lsd latch on*/

    for(;;)   /*Start of loop*/
```
```

```
{
    out2_til311(CNT);      /*Output to TIL311*/
    CNT++;                 /*Increment count*/
    if(CNT == 100)CNT=0;   /*Back to 0 if 100*/
    wait_a_second();       /*Wait a second*/
    }
}
```

Figure 3.16.

Program listing of Project 7

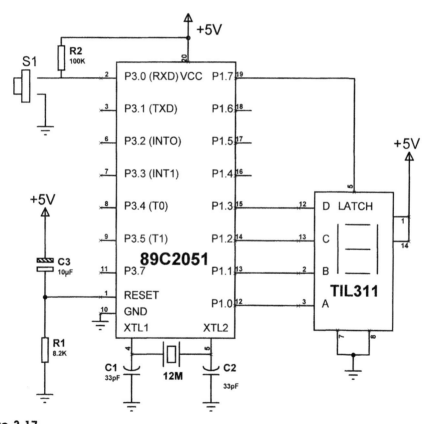

Figure 3.17.

Circuit diagram of Project 8

the button is not pressed, a count is incremented between 1 and 6. When the push-button is pressed, the current value of the count is sent to the display by calling the function *out_til311*. The above process continues after about 2 seconds delay.

The following PDL describes the functions of the program:

Main program

START
    Set latch to 1
    DO FOREVER
        IF button is pressed **THEN**
            Call function out_til311 with value of count
            Delay 2 seconds
            Clear the display
        ELSE
            Increment count
            IF count = 7 **THEN**
                Count = 1
            ENDIF
        ENDIF
    ENDDO
END

Function out_til311

Input: Count
Output: None

START
    Set top bit of count
    Send count to port 1
    Set LATCH to LOW
    Set LATCH to HIGH
END

Program Listing

The full program listing is shown in Fig. 3.18 (see pp. 58 and 59). Display latch (variable *latch*) is assigned to bit 7 of port 1 using the instruction *sbit*. Similarly, the push-button is assigned to bit 0 of port 3 and is named *button*. Variable *DICE* stores the random dice values.

The display latch is set to 1 to avoid any unwanted write to the display and the endless loop starts with the *for* statement. When the button is pressed (*button* = 0), the current value of *DICE* is sent to function *out_til311* which displays the value. After 2 seconds delay the display is cleared and the program loop continues from the beginning.

Light Projects 57

If the button is not pressed (*button* = 1), the value of *DICE* is incremented until it is 7 and then set back to 1.

Components Required

In addition to the components used for the basic microcontroller circuit, the following components will be required:

Display   TIL311 type hexadecimal display
R2   100K, 0.125 W resistor

PROJECT 9 - 7 Segment Display Driver

Function

This project shows how a 7 segment display can be interfaced to a microcontroller. In this project, a 7 segment display is connected to port 1 of the microcontroller and a program is written to count up from 0 to 9 and display the data on the 7 segment display. The program runs in an endless loop and a 1 second delay is used between each output.

Circuit Diagram

Seven segment displays are used in many industrial and commercial applications. Basically the display consists of seven segments of LEDs, connected either as common anode or common cathode. In a common-anode display the anodes of all the LED segments are connected together. Similarly, all the cathodes are connected together in a common-cathode display. Segments in a 7 segment display are identified by giving them letters from a to g as shown in Fig. 3.19.

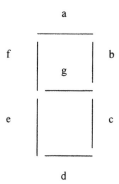

Figure 3.19.

Segments of a 7 segment display

/***
PROJECT: PROJECT 8
FILE: PROJ8.C
DATE: August 1999
PROCESSOR: AT89205 1

This is a dice simulator project. A TIL311 type hexadecimal alphanumeric display is
connected to port 1 of the microcontroller.

When a push-button, connected to bit 0 of port 3, is depressed, a random number
between 1 and 6 is displayed on the hexadecimal display. After about 2 seconds the
displayed is cleared and the user can throw the dice again.
***/
#include <AT892051.h>

sbit latch = P1^7;   /*Bit P1.7 is the latch*/
sbit button = P3^0;   /*Bit P3.0 is push-button*/

/* Function to delay about a second */
void wait_a_second()
{
   unsigned int x;
   for(x=0;x<33000;x++);
}

/* Function to display data on a TIL311 display */
void out_til311(unsigned char x)
{
   P1=x | 0x80;   /*Send data*/
   latch=0;   /*Latch the data*/
   latch=1;   /*Set latch on*/
}

/* Start of main program */
main()
{
   unsigned char DICE;

   latch=1;   /*Set latch on*/

   for(;;)   /*Start of loop*/
   {
      if(button == 0)   /*Button pressed*/
      {
         out_til311(DICE); /*Display DICE*/
         wait_a_second();   /*Wait 2 seconds..*/
         wait_a_second();

```
      out_til311(0);              /*Clear display*/
   }
   else
   {
      DICE++;                     /*Inc DICE*/
      if(DICE == 7)DICE=1;        /*Set to 1 if 7*/
   }
  }
}
```

Figure 3.18.

Program listing of Project 8

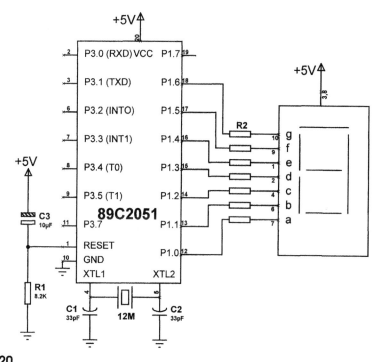

Figure 3.20.

Circuit diagram of Project 9

Required characters are generated by turning on the appropriate LED segments. Table 3.3 shows the segments that should be turned on to generate the decimal numbers 0 to 9. A 1 in the table corresponds to the segment being on.

The circuit diagram of Project 9 is shown in Fig. 3.20. A common-anode type display is used in this project. The anode pins (3 and 8) are connected to +5 V.

Table 3.3	
Number	g f e d c b a
0	0 1 1 1 1 1 1
1	0 0 0 0 1 1 0
2	1 0 1 1 0 1 1
3	1 0 0 1 1 1 1
4	1 1 0 0 1 1 0
5	1 1 0 1 1 0 1
6	1 1 1 1 1 0 0
7	0 0 0 0 1 1 1
8	1 1 1 1 1 1 1
9	1 1 0 0 1 1 1

Segments a to g are connected to port 1 of the microcontroller via 470 Ω current limiting resistors. Segment a is connected to bit 0 of port 1, segment b to bit 1 of port 1, segment c to bit 2 of port 1 and so on.

Program Description

A bit map table has been created which shows the segments to be turned on and the corresponding hexadecimal numbers that should be sent to the display in order to display the required numbers, as shown in Table 3.4 (in this table x is not used but included in the table so that the hexadecimal numbers can be derived easily as two 4-bit nibbles).

The following PDL describes the functions of the program:

START
    Initialize count and bit pattern array
    DO FOREVER
        Get bit pattern corresponding to count
        Output bit pattern to port 1
        Delay 1 second
        Increment count
        IF count = 10 **THEN**
            Count = 0
        ENDIF
    ENDDO
END

Table 3.4 Segments and corresponding bit patterns

Number	x g f e d c b a	Hex
0	0 0 1 1 1 1 1 1	3F
1	0 0 0 0 0 1 1 0	06
2	0 1 0 1 1 0 1 1	5B
3	0 1 0 0 1 1 1 1	4F
4	0 1 1 0 0 1 1 0	66
5	0 1 1 0 1 1 0 1	6D
6	0 1 1 1 1 1 0 0	7C
7	0 0 0 0 0 1 1 1	07
8	0 1 1 1 1 1 1 1	7F
9	0 1 1 0 0 1 1 1	67

Program Listing

The full program listing is shown in Fig. 3.21. Variable *LED* is initialized to 0. The 7 segment bit pattern is loaded into array *LED_ARRAY*. The endless loop is started with the *for* statement. Data is sent to the display by indexing the *LED_ARRAY* with the number to be displayed. The data is complemented before it is output since the output ports are sourcing current and a segment is turned on when the corresponding output bit is at logic 0. After a 1 second delay, the variable *LED* is incremented by one, ready for the next display. When variable *LED* reaches 10 it is reset back to 0.

The following data is displayed by the 7 segment display:

0 1 2 3 4 5 6 7 8 9 0 1 2 . . .

Components Required

In addition to the components used for the basic microcontroller circuit, the following components will be required:

R2 470 Ω, 0.125 W resistors (8 off)
Display 7 segment common-anode display

62 Microcontroller Projects in C for the 8051

```
/******************************************************************************
PROJECT:               PROJECT 9
FILE:                  PROJ9.C
DATE:                  August 1999
PROCESSOR:             AT892051

This is a 7 segment display interface project. The display is connected to port 1 of the
microcontroller and counts up from 0 to 9 with 1 second delay between each count.
******************************************************************************/

#include <AT892051.h>

/* Function to delay about a second */
void wait_a_second()
{
   unsigned int x;
   for(x=0;x<33000;x++);
}

/* Start of main program */
main()
{
   int LED=0;                    /*Initialize to 0*/
   int LED_ARRAY(10)=
      {0x3F,0x06,0x5B,0x4F,0x66,0x6D,0x7C,0x07, 0x7F,0x67
      };

   for(;;)                       /*Start of loop*/
   {
      P1=~LED_ARRAY(LED);        /*LED on*/
      wait_a_second();           /*Wait a sec*/
      LED++;                     /*Inc count*/
      if(LED == 10)LED=0;        /*Set to 0*/
   }
}
```

Figure 3.21.
Program listing of Project 9

PROJECT 10 – Four-digit LED Display Interface

Function

This project shows how a 4-digit display can be interfaced to a microcontroller. The display we shall be using in this project is the TSM5X34 series 0.3″ 4-digit

Light Projects

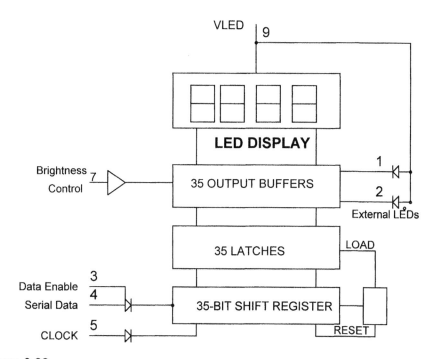

Figure 3.22.
TSM5034 4-digit display

display with on-board driver. This display can be used in many microcontroller-based applications, including digital clocks, thermometer circuits, instrument readouts, counters, voltmeters and so on. In this project we shall design a 4-digit up-counter which counts from 0 to 9999. We shall be using this display in some of our other projects as a visual readout device.

TSM5X34 Series Displays

The TSM5X34 is a 0.3" 4-digit display with on-board serial data input (Fig. 3.22). Serial data transfer from a microcontroller to the display is accomplished with three signals: serial data, data enable, and clock. The data format consists of a leading '1', followed by 35 data bits, each bit corresponding to the segments to be turned on in the display. All of the four digits are programmed at the same time by sending 35 bits of serial data to the display. The clock input is pulsed after each data is sent. The enable input should be at logic 0 to enable programming of the device.

There are mainly two versions of the TSM series of displays. TSM5xxx devices can drive two external LEDs and TSM6xxx series devices incorporate the

Table 3.5 Bit patterns for each segment of the display

Bit	Digit	Segment	Bit	Digit	Segment
1	1	A	18	3	B
2	1	B	19	3	C
3	1	C	20	3	D
4	1	D	21	3	E
5	1	E	22	3	F
6	1	F	23	3	G
7	1	G	24	3	Dp
8	1	Dp	25	4	A
9	2	A	26	4	B
10	2	B	27	4	C
11	2	C	28	4	D
12	2	D	29	4	E
13	2	E	30	4	F
14	2	F	31	4	G
15	2	G	32	4	Dp
16	2	Dp	33	–	LED1
17	3	A	34	–	LED2

colon character as part of the display. In the TSM5xxx series, TSM5034 emits red light, TSM5234 emits green light, and TSM5734 emits high efficiency red light. In this project we shall be using the popular TSM5034 type display.

Thirty-five bits of data should be sent to the display following a '1' start bit. Table 3.5 shows the bit patterns for each segment of the display.

As an example, suppose that we want to turn on segment B of digit 1, segment B of digit 2, segments C and D of digit 3, and segments A, B, and C of digit 4, and we are not connecting any external LEDs to the display, and the decimal points should be off. The bit pattern shown in Fig. 3.23 should then be sent to the display (each bit should be clocked by sending a clock pulse).

Light Projects

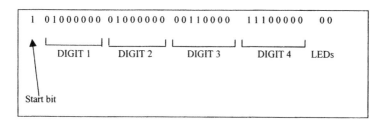

Figure 3.23.
Bit pattern for the example

Table 3.6 Numbers and corresponding TSM5034 bit patterns		
Number	A B C D E F G Dp	Hex code
0	1 1 1 1 1 1 0 0	FC
1	0 1 1 0 0 0 0 0	60
2	1 1 0 1 1 0 1 0	DA
3	1 1 1 1 0 0 1 0	F2
4	0 1 1 0 0 1 1 0	66
5	1 0 1 1 0 1 1 0	B6
6	1 0 1 1 1 1 1 0	BE
7	1 1 1 0 0 0 0 0	E0
8	1 1 1 1 1 1 1 0	FE
9	1 1 1 1 0 1 1 0	F6

The easiest way of controlling the TSM display is to create a table of bit patterns for each decimal digit 0 to 9. If we assume that segment A is the most significant bit in this table, for a given digit we can read the required bit pattern from the table and then send bits to the display by shifting the bits left, one bit at a time for each digit. The total number of data bits sent will be $8 \times 4 = 32$ bits for 4 digits, 2 bits for the two LEDs, making a total of 34. In addition, we have to send a start bit, making an overall total of 35 bits.

Table 3.6 shows the relationship between the decimal numbers 0 to 9, the corresponding TSM5034 bit patterns, and the corresponding values in hexadecimal.

As an example, if we want to display the decimal number 2367, we have to send the following bit pattern:

```
1          Start bit
11011010   Hexadecimal DA for decimal digit 2
11110010   Hexadecimal F2 for decimal digit 3
10111110   Hexadecimal BE for decimal digit 6
11100000   Hexadecimal E0 for decimal digit 7
0          turn off LED 1
0          turn off LED 2
```

It will therefore be necessary to convert a given number into four decimal digits and then use the above technique to display it.

A required digit can be totally blanked by sending all zeros for that digit. This could be useful when it is required to blank the leading digits instead of displaying zeros if the number to be displayed is less than four digits.

Circuit Diagram

The circuit diagram of Project 10 is shown in Fig. 3.24. The circuit is very simple. Bit 6 of port 1 is connected to the clock input of the display. Similarly, bit 7 of port 1 is connected to the data input of the display. Display pin 7 is the brightness control input and it should be connected to the supply voltage via a suitable resistor. A 0.01 µF capacitor is recommended by the manufacturer as it stops any oscillations. VDD and VLED should normally be connected to +5 V, although VLED can also be connected to a smaller voltage for lower power consumption. The enable input (pin 3) is connected to the ground. In multiple display operations it is necessary to control the enable input of each device individually so that data and clock can be routed to the required display.

Program Description

The program is a simple 4-digit decimal counter. A counter is initialized to 0 and the display is cleared at the beginning of the program. The counter is then displayed and incremented in an endless loop. The following PDL describes the functions of the program:

Light Projects

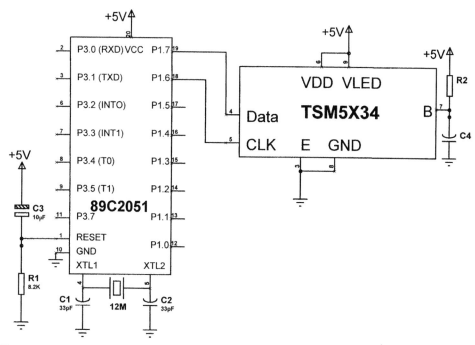

Figure 3.24.
Circuit diagram of Project 10

Main program

START
    Initialize counter to 0
    Call function clear_display
    DO FOREVER
        Call function display_all with counter
        Delay
        Increment count
    ENDDO
END

Function clear_display

Input: None
Output: None

START
    Set display data to 1 (start bit)
    Call function send_clock to send a clock pulse

Set display data to 0
Send 35 clock pulses
END

Function send_clock

Input: None
Output: None

START
Set display clock input to 1
Set display clock input to 0
END

Function display_all

Input: Count
Output: None

START
Convert data into 4 decimal digits
Call function display_digit to display Digit 1
Call function display_digit to display Digit 2
Call function display_digit to display Digit 3
Call function display_digit to display Digit 4
END

Function display_digit

Input: Digit value
Output: None

START
Get bit map of the digit to be displayed
Get top bit of the bit pattern
IF top bit = 0 **THEN**
Send 0 to the display data input
ELSE
Send 1 to the display data input
ENDIF
Send a clock pulse to the display
Shift bit map left by 1 bit
END

Program Listing

The full program listing is shown in Fig. 3.25. Variable *LED* is initialized to 0 at the beginning of the main program. Function *clear_display* is then called to clear all digits of the display. The endless loop starts with the *for* statement. Inside this loop, function *display_all* is called to display the value of variable *LED* as four digits. digits on the TSM5034. The value of variable *LED* is then incremented and the loop is repeated forever.

Function *display_all* receives the number to be displayed as its argument (*n* in this case). This number can range from 0 to 9999. First of all this number is converted into four decimal digits and these digits are stored in integer variables *first*, *second*, *third*, and *fourth*, where first is digit 1 and *fourth* is digit 4. Function *display_digit* is then called to display the digit values. This function stores the bit map of the display in array *LED_ARRAY*. The bit map corresponding to the number to be displayed is obtained by the statement $n = LED\_ARRAY[x]$. The top bit of this bit map is then examined. If the top bit is a 1 then a 1 is sent to the data input of the display, otherwise a 0 is sent to the data input of the display. The display is then clocked by calling the function *send_clock*, which sends a single clock pulse to the display. The bit map data is then shifted left using the shift operator '≪' so that the second bit can be tested and sent to the display. This is repeated for all 8 bits of the bit map.

Blanking Leading Zeros

The program listing given in Fig. 3.25 displays the data as a 4-digit number with leading zeros. For example, number 27 is displayed as 0027. There are many applications where we may want to blank the leading zeros. It is possible to blank the leading zeros by setting all segments of the leading zero digits to the off state. This is done in the program listing shown in Fig. 3.26. Here, a new bit map 0 is introduced into array *LED_ARRAY* and the array dimension is increased to 11. The new bit map is indexed with number 10. Function *display_all* is changed so that blanks are displayed instead of leading zeros when the number of digits is less than 4. For example, if the number to be displayed is less than 1000, the first digit is displayed by using the bit map defined by *LED_ARRAY[10]*, which is 0, i.e. all the segments of the digit are set to 0. Similarly, if the number to be displayed is less than 100, the second (and first) digit is displayed with the segments turned off.

General Display Program

There may be some applications where we may need to show leading zeros and also the decimal points. The program listing given in Fig. 3.27 enables both the leading zeros and the decimal points to be optionally displayed. Function

```
/******************************************************************************
PROJECT:                    PROJECT 10
FILE:                       PROJ10.C
DATE:                       August 1999
PROCESSOR:                  AT892051

This is a TSM5034-based 4-digit display interface project. The display counts up from 0 to
9999 with about 1 second delay between each count.
******************************************************************************/
#include <AT892051.h>

sbit DISPLAY_CLOCK=P1^6;
sbit DISPLAY_DATA =P1^7;

/* Function to delay about a second */
void wait_a_second()
{
  unsigned int x;
  for(x=0;x<33000;x++);
}

/* Function to send a clock pulse to the display */
void send_clock()
{
  DISPLAY_CLOCK=1;
  DISPLAY_CLOCK=0;
}

/* Function to display a single digit */
void display_digit(int x)
{
  unsigned char LED_ARRAY(10)=
  {
  0xFC,0x60,0xDA,0xF2,0x66,0xB6,0xBE, 0xE0,0xFE,0xF6
  };

  unsigned char n,top_bit,i;

  n=LED_ARRAY(x);
  for(i=1;i<=8;i++)
  {
    top_bit=n & 0x80;        /*Get top bit*/
    if(top_bit != 0)
      DISPLAY_DATA=1;
    else
      DISPLAY_DATA=0;
```

Light Projects

```c
      send_clock();
      n=n << 1;                   /*Shift left by 1 digit*/
   }
}

/* Function to display all 4 digits*/
void display_all(int n)
{
   int r,first,second,third,fourth;

   first=n/1000;
   r=n-1000*first;
   second=r/100;
   r=r-100*second;
   third=r/10;
   fourth=r-third*10;

   DISPLAY_DATA=1;
   send_clock();
   display_digit(first);         /* display digit 1*/
   display_digit(second);        /*display digit 2*/
   display_digit(third);         /*display digit 3*/
   display_digit(fourth);        /*display digit 4*/

   DISPLAY_DATA=0;
   send_clock();                 /*35 clks required*/
   send_clock();
   send_clock();
}
/* Function to clear the display */
void clear_display()
{
  int i;
  DISPLAY_DATA=0;
  DISPLAY_CLOCK=0;
  DISPLAY_DATA=1;
  send_clock();
  DISPLAY_DATA=0;
  for(i=1;i<=35;i++)send_clock();
}

/* Start of main program */
main()
{
   int LED=0;                    /*initialize to 0*/
```

```
    clear_display();              /*Clear display*/

    for(;;)                       /*Start of loop*/
    {
        display_all(LED);
        wait_a_second();          /*Wait a second*/
        LED++;                    /*Increment count*/
    }
}
```

Figure 3.25.
Program listing of Project 10

display_all displays an integer number between 0 and 9999. This function is called with the following arguments:

display_all(n,lz,dp1,dp2,dp3,dp4)

where:

n   is the number to be displayed
lz   is the leading zero blanking flag. If lz=0, data is displayed with leading zeros. If lz=1, data is displayed with leading zeros blanked.
dp1 to dp4   these are the decimal point enable bits for digits 1 to 4 respectively. For example, if dp1=0 then the decimal point of digit 1 is not displayed. If on the other hand, dp1=1 then the decimal point of digit 1 is displayed.

For example, the function call:

display_all(124,0,0,1,0)

will display the following data:

012.4

i.e. leading zeros are enabled and a decimal point is inserted after digit 3.

Components Required

In addition to the components used for the basic microcontroller circuit, the following components will be required for this project:

Light Projects   73

```
/******************************************************************************
PROJECT:                PROJECT 10
FILE:                   PROJ10-1.C
DATE:                   August 1999
PROCESSOR:              AT892051

This is a 7 segment display interface project. The display counts up from 0 to 9 with about
1 second delay between each count.

This program blanks the unused leading digits.
******************************************************************************/
#include <AT892051.h>

sbit DISPLAY_CLOCK=P1^6;
sbit DISPLAY_DATA =P1^7;

/* Function to delay about a second */
void wait_a_second()
{
   unsigned int x;
   for(x=0;x<33000;x++);
}

/* Function to send a clock pulse to the display */
void send_clock()
{
   DISPLAY_CLOCK=1;
   DISPLAY_CLOCK=0;
}

/* Function to display a single digit */
void display_digit(int x)
{
   unsigned char LED_ARRAY(11)=
   {
      0xFC,0x60,0xDA,0xF2,0x66,0xB6,0xBE,0xE0,0xFE,0xF6,0
   };
   unsigned char n,top_bit,i;

   n=LED_ARRAY(x);
   for(i=1;i<=8;i++)
   {
      top_bit=n & 0x80;           /*Get top bit*/
      if(top_bit != 0)
         DISPLAY_DATA=1;
      else
```

```
        DISPLAY_DATA=0;
    send_clock();
    n=n << 1;                   /*Shift left by 1 digit*/
  }
}


/* Function to display all 4 digits */
void display_all(int n)
{
    int r,first,second,third,fourth;

    first=n/1000;
    r=n-1000*first;
    second=r/100;
    r=r-100*second;
    third=r/10;
    fourth=r-third*10;

    DISPLAY_DATA=1;
    send_clock();

    if(n < 1000)                /*Blank leading zero*/
        display_digit(10);
    else
        display_digit(first);
    if(n < 100)
        display_digit(10);
    else
        display_digit(second);
    if(n < 10)
        display_digit(10);
    else
        display_digit(third);
    display_digit(fourth);

    DISPLAY_DATA=0;
    send_clock();               /*35 clks required*/
    send_clock();
    send_clock();
}

/* Function to clear the display */
void clear_display()
{
```

Light Projects

```
    int i;
    DISPLAY_DATA=0;
    DISPLAY_CLOCK=0;
    DISPLAY_DATA=1;
    send_clock();
    DISPLAY_DATA=0;
    for(i=1;i<=35;i++)send_clock();
}

/* Start of main program */
main()
{
    int LED=0;              /*initialize to 0*/
    clear_display();        /*Clear display*/

    for(;;)                 /*Start of loop*/
    {
        display_all(LED);
        wait_a_second();    /*Wait a second*/
        LED++;              /*Increment count*/
    }
}
```

Figure 3.26

Program listing of Project 10 with leading zeros blanked

R2	8.2K, 0.125 W resistor
C4	0.01 µF capacitor
Display	TSM5034

PROJECT 11 – Interrupt Driven Event Counter with 4-digit LED Display

Function

This project shows how the external interrupt input of the microcontroller can be programmed using the C language. The project is a simple interrupt-based event counter which can count external events from 0 to 9999. A TSM5034 type 4-digit display is connected to port 1 of the microcontroller. Bit 0 of port 3 is connected to a push-button switch S1 which is used to clear the display whenever required. External interrupt input INT0 (pin 6) of the

```
/****************************************************************
PROJECT:              PROJECT 10
FILE:                 PROJ10-2.C
DATE:                 August 1999
PROCESSOR:            AT892051

This is a 4-digit TSM5034 display interface project. The display counts up from 0 to 9999
with about 1 second delay between each count.

This program shows the leading zeros with decimal points between digit 2 and 3.
****************************************************************/
#include <AT892051.h>

sbit DISPLAY_CLOCK=P1^6;
sbit DISPLAY_DATA =P1^7;

/* Function to delay about a second */
void wait_a_second()
{
  unsigned int x;
  for(x=0;x<33000;x++);
}

/* Function to send a clock pulse to the display */
void send_clock()
{
  DISPLAY_CLOCK=1;
  DISPLAY_CLOCK=0;
}

/* Function to display a single digit */
void display_digit(int x,char dp)
{
  unsigned char LED_ARRAY[11]=
  {
    0xFC,0x60,0xDA,0xF2,0x66,0xB6,0xBE,0xE0,0xFE,0xF6,0
  };
  unsigned char n,top_bit,i;
  n=LED_ARRAY[x] | dp;       /*Set decimal point*/
  for(i=1;i<=8;i++)
  {
    top_bit=n & 0x80;        /*Get top bit*/
    if(top_bit != 0)
      DISPLAY_DATA=1;
    else
```

```
      DISPLAY_DATA=0;
    send_clock();
    n=n << 1;
  }
}

/* Function to display all 4 digits */
void display_all(int n,char lz,char dp1,char dp2,char dp3,char dp4)
{
   int r,first,second,third,fourth;

   first=n/1000;
   r=n-1000*first;
   second=r/100;
   r=r-100*second;
   third=r/10;
   fourth=r-third*10;

   DISPLAY_DATA=1;
   send_clock();
   /* Blank leading zeros */
   if(n < 1000 && lz == 1)
      display_digit(10,0);
   else
      display_digit(first,dp1);
   if(n < 100 && lz == 1)
      display_digit(10,0);
   else
      display_digit(second,dp2);
   if(n < 10 && lz == 1)
      display_digit(10,0);
   else
      display_digit(third,dp3);
   if(lz == 1)
      display_digit(fourth,0);
   else
      display_digit(fourth,dp4);

   DISPLAY_DATA=0;
   send_clock();
   send_clock();
   send_clock();
}

/* This function clears the display */
```

```c
void clear_display()
{
  int i;
  DISPLAY_DATA=0;
  DISPLAY_CLOCK=0;
  DISPLAY_DATA=1;
  send_clock();
  DISPLAY_DATA=0;
  for(i=1;i<=35;i++)send_clock();
}

/* Start of main program */
main()
{
  int LED=0;              /*initialize to 0*/
  clear_display();        /*Clear display*/

  for(;;)                 /*Start of loop*/
  {
    display_all(LED,0,0,1,0,0);
    wait_a_second();      /*Wait a second*/
    LED++;                /*Increment count*/
  }
}
```

Figure 3.27.

Program listing of a more general display program

microcontroller is used as an edge-triggered event input. An external event occurs when INT0 is clocked from 1 to 0.

Circuit Diagram

Figure 3.28 shows a block diagram of the hardware. The push-button switch is the reset input. External events are falling edge triggered. A TSM5034 display shows the event count at any time.

The complete circuit diagram of this project is shown in Fig. 3.29. TSM5034 is connected to port 1 of the microcontroller. The clock input is connected to bit 6 of port 1 and the data input is connected to bit 7 of port 1. Bit 0 of port 3 is connected to the event reset switch S1. This input is normally held at logic 1 with the pull-up resistor R3. When the switch is pressed the pin goes to logic 0 which can be detected by the software. External interrupt input INT0 is used as the event counter input. This pin is normally held at logic 1 with the pull-up resistor R4. An external event occurs when this pin is clocked to 0. This

Light Projects 79

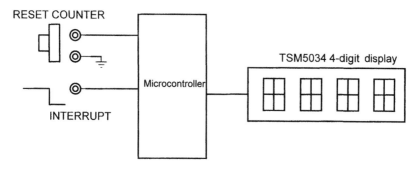

Figure 3.28.
Block diagram of the event counter

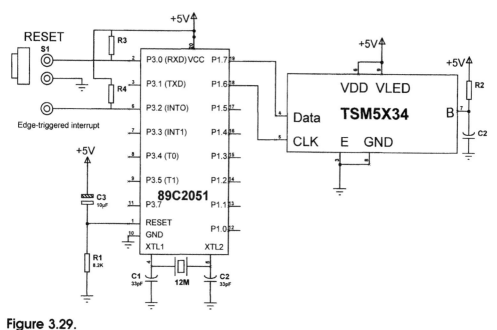

Figure 3.29.
Circuit diagram of Project 11

generates an interrupt in the software which increments the count and displays the total number of events occurred.

Program Description

The program first initializes the interrupt registers of the microcontroller so that external interrupts on pin INT0 can be detected. An endless loop is then formed with a *for* statement. Inside this loop the reset input (bit 0 of port 3) is checked and when the reset switch is pressed, the counter is cleared to zero. The interrupt service routine simply increments the current event count and displays the result.

The following PDL describes the functions of the program:

Main program

START
    Clear display
    Initialize External interrupt INT0
    DO FOREVER
        IF RESET switch is pressed **THEN**
            Clear event counter
            Clear display
        ENDIF
    ENDDO
END

Interrupt service routine

START
    Increment event counter
    Display event counter
END

The display part of the program is the same as in Project 10 and is not described here again.

Program Listing

The full program listing is shown in Fig. 3.30. Variable *DISPLAY_CLOCK* is the clock input of the display and is assigned to bit 6 of port 1. *DISPLAY_DATA* is the data input of the display and is assigned to bit 7 of port 1. Bit 0 of port 3 is assigned to variable *RESET_COUNTER*. Variable *EVENT* is used as the event counter. The program first of all clears the display. The interrupt registers of the microcontroller are then programmed. Statement $IT0 = 1$ sets external interrupt input INT0 so that interrupts can be recognized on the falling edge (1 to 0) of this pin. Statement $EX0 = 1$ enables external interrupt INT0. Statement $EA = 1$ enables interrupts so that they can be accepted by the microcontroller. Inside the endless loop the RESET input is checked. If the user presses RESET (i.e. $RESET\_COUNTER = 0$), the counter value *EVENT* is reset to zero and the display is cleared.

Light Projects **81**

```
/****************************************************************************
PROJECT:                    PROJECT 11
FILE:                       PROJ11.C
DATE:                       August 1999
PROCESSOR:                  AT892051

This is a 4-digit TSM5034 display interface project. The display counts up from 0 to 9999
with about 1 second delay between each count.

This program shows the leading zeros with decimal points between digit 2 and 3.
****************************************************************************/
#include <AT892051.h>

sbit DISPLAY_CLOCK=P1^6;
sbit DISPLAY_DATA =P1^7;
sbit RESET_COUNTER=P3^0;

int EVENT=0;                      /*initialize to 0*/

/* Function to send a clock pulse to the display */
void send_clock()
{
  DISPLAY_CLOCK=1;
  DISPLAY_CLOCK=0;
}

/* Function to display a digit */
void display_digit(int x,char dp)
{
  unsigned char LED_ARRAY[11]=
  {
    0xFC,0x60,0xDA,0xF2,0x66,0xB6,0xBE,0xE0,0xFE,0xF6,0
  };
  unsigned char n,top_bit,i;

  n=LED_ARRAY(x) | dp;            /*decimal point*/
  for(i=1;i<=8;i++)
  {
    top_bit=n & 0x80;             /*Get top bit*/
    if(top_bit != 0)
      DISPLAY_DATA=1;
    else
      DISPLAY_DATA=0;
    send_clock();
    n=n << 1;                     /*Shift by 1*/
  }
}
```

```c
/* Function to display all 4 digits */
void display_all(int n,char lz,char dp1,char dp2,char dp3,char dp4)
{
   int r,first,second,third,fourth;

   first=n/1000;
   r=n-1000*first;
   second=r/100;
   r=r-100*second;
   third=r/10;
   fourth=r-third*10;

   DISPLAY_DATA=1;
   send_clock();

   if(n < 1000 && lz == 1)        /*Blank leading zero*/
      display_digit(10,0);
   else
      display_digit(first,dp1);
   if(n < 100 && lz == 1)
      display_digit(10,0);
   else
      display_digit(second,dp2);
   if(n < 10 && lz == 1)
      display_digit(10,0);
   else
      display_digit(third,dp3);
   if(lz == 1)
      display_digit(fourth,0);
   else
      display_digit(fourth,dp4);

   DISPLAY_DATA=0;
   send_clock();
   send_clock();
   send_clock();
}

/* Function to clear the display */
void clear_display()
{
   int i;
   DISPLAY_DATA=0;
   DISPLAY_CLOCK=0;
   DISPLAY_DATA=1;
```

```
    send_clock();
    DISPLAY_DATA=0;
    for(i=1;i<=35;i++)send_clock();
}

/* External interrupt INT0 service routine */
cnt() interrupt 0
{
    EVENT++;
    display_all(EVENT,1,0,0,0,0);
}

/* Start of main program */
main()
{
    clear_display();            /*Clear display*/
    IT0=1;                      /*Interrupt on falling-edge*/
    EX0=1;                      /*Enable interrupt INT0*/
    EA=1;                       /*Enable interrupts*/

    for(;;)                     /*Start of endless loop*/
    {
        if(RESET_COUNTER == 0)
        {
            EVENT=0;
            clear_display();
        }
    }
}
```

Figure 3.30.

Program listing of Project 11

The interrupt service routine is declared by the function *cnt() interrupt 0*, where 0 is the interrupt number. 89C2051 interrupt numbers are defined as shown in Table 3.7.

Whenever input INT0 goes from logic 1 to 0 an external interrupt is generated and the program jumps to interrupt service routine declared by function *cnt() interrupt 0*. This routine increments the event counter and displays the result on the TSM5034. The displayed value is thus equal to the total number of events on pin INT0.

Components Required

In addition to the standard components used by the microcontroller, the following components will be required:

Table 3.7 Interrupt numbers

Interrupt No.	Description
0	External interrupt 0
1	Timer 0 interrupt
2	External interrupt 1
3	Timer 1 interrupt
4	Serial port interrupt

Display   TSM5034 4-digit display
R2   8.2K, 0.125 W resistor
R3, R4   100K, 9.125 W resistors
C4   0.01 µF capacitor
S1   push-button switch

CHAPTER 4

SOUND PROJECTS

In this chapter we shall be looking at how we can interface our microcontroller to sound generating devices. Sound projects are based on audible devices and these devices have many applications in electronics, ranging from warning devices, burglar alarms, speech processing applications, electronic organs and so on.

Electronic sound generation requires an electronic audible device. There are several such devices available:

- *Piezo sounders*: these devices operate by an external DC source. An internal oscillator applies an AC signal to a piezo substrate and this causes alternating deformation of the disc, producing sound output. These devices require about 8 to 20 mA current and generate a sound output of 80 to 100 dBA, at a distance of approximately 30 cm. The frequency response of these devices is in a narrow band, generally in the region 3 to 5 KHz. Piezo sounders usually emit a single tone but some models can emit two or more tones and can also provide pulsed tone outputs. Piezo sounders operate over a wide DC voltage range and as a result of this, they are widely used in small portable electronic equipment.
- *Buzzers*: these are mechanical devices which produce sound via a magnetized arm repeatedly striking a diaphragm. These devices operate with a DC voltage and the current requirement is small, generally in the region of 10 mA. Buzzers generate a 'buzzing' noise (single tone) in the frequency range 300 to 500 Hz. Buzzers are small devices and they can be either panel mounted or PCB mounted.
- *Sounders*: these audible devices generally operate with a DC voltage in the range 3 to 24 V. The current requirement is around 15 mA. The sound output of sounders is single tone at 3 KHz or less, with 80 to 85 dBA at a distance of 30 cm.
- *Transducers*: these devices generally operate with a small DC voltage (around 3 V) and require external drive circuitry. Sound output is 85 dBA or more at a distance of 30 cm. The resonant frequency of transducers is 3 KHz or less. These devices are usually used as mini speakers in PCB mounted applications.

- *Coil type*: these devices operate by a coil attracting and repelling a magnetized diaphragm. The principle of operation is the same as a loudspeaker and in fact these are tiny speakers. An external drive circuit is required to generate sound. Coil type audible devices are generally used when multitone sound or speech is required.

In this chapter we shall be interfacing our microcontroller to simple buzzers and also to more complex audible devices.

PROJECT 12 – Simple Buzzer Interface

Function

This project shows how we can interface our microcontroller to a buzzer. When a push-button switch is pressed (e.g. simulating a burglary), the buzzer will turn on and off 30 times and then stop.

Circuit Diagram

The circuit diagram of this project is shown in Fig. 4.1. Bit 7 of port 1 is connected directly to a small buzzer. This type of connection is possible if the current requirement of the buzzer is not more than about 20 mA. The port output is in current source mode so that the buzzer will turn on when the port output is at logic LOW (0 V). Bit 0 of port 3 is connected to a push-button switch which is normally held at logic HIGH by a pull-up resistor.

Program Description

The buzzer is initially turned OFF. The push-button switch is then checked and when the switch is pressed, the buzzer is turned on and off 30 times, with a 1 second delay between each output. The following PDL describes the functions of the program:

```
START
        Turn OFF buzzer
        DO FOREVER
                IF push-button switch pressed THEN
                        DO 30 times
                                turn ON buzzer
                                delay
                                turn OFF buzzer
                        ENDDO
                ENDIF
        ENDDO
END
```

Sound Projects **87**

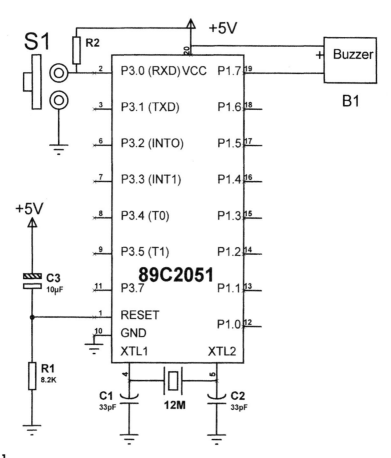

Figure 4.1.
Circuit diagram of Project 12

Program Listing

The program listing is given in Fig. 4.2. Variable *BUZZER* is defined as bit 7 of port 1. Similarly, variable *PUSH_BUTTON* is defined as bit 0 of port 3. When the program starts, *BUZZER* is set to OFF, where OFF is defined as logic HIGH. The state of the *PUSH_BUTTON* switch is then checked continuously in a loop. When the switch is pressed (*PUSH_BUTTON* = 0), a *for* loop is set to repeat 30 times. Inside this loop the buzzer is turned ON and OFF with a 1 second delay between each output.

Using Higher Current Buzzers

The buzzer used in Fig. 4.1 is assumed to draw not more than 20 mA and thus we can connect the buzzer directly to the microcontroller. For buzzers that

Microcontroller Projects in C for the 8051

```
/********************************************************************
PROJECT:                PROJECT 12
FILE:                   PROJ12.C
DATE:                   August 1999
PROCESSOR:              AT892051

This is a simple buzzer project. The buzzer sounds on and off for 30 seconds when a push-
button switch is pressed.

The buzzer is connected to bit 7 of port 1 directly and the buzzer is ON when the output of
the port is at logic LOW, i.e. when the output port is sourcing current. The push-button
switch is connected to bit 0 of port 3.
********************************************************************/
#include <AT892051.h>

sbit BUZZER=P1^7;
sbit PUSH_BUTTON =P3^0;

#define ON 0
#define OFF 1

/* Function to delay about a second */
void wait_a_second()
{
    unsigned int x;
    for(x=0;x<33000;x++);
}

/* Start of main program */
main()
{
    int i;

    BUZZER=OFF;                         /*Turn buzzer off*/
    for(;;)                             /*endless loop*/
    {
        while(PUSH_BUTTON == 1)  /*wait for push-button*/
        {
        }
        for(i=1;i<=30;i++)              /*do 30 times*/
        {
            BUZZER=ON;                  /*turn on buzzer*/
            wait_a_second();            /*delay a second*/
            BUZZER=OFF;                 /*turn off buzzer*/
            wait_a_second();            /*delay a second*/
        }
    }
}
```

Figure 4.2.
Program listing of Project 12

Sound Projects

operate with higher currents it will be necessary to use a circuit similar to Fig. 4.3. In this circuit a MOSFET n-channel transistor is used as a switch. The buzzer is connected to the drain (D) input and the gate (G) input is driven directly from the microcontroller. The current drawn by the gate input is practically zero. A VN66AFD type MOSFET can be used to switch currents up to about 2 A.

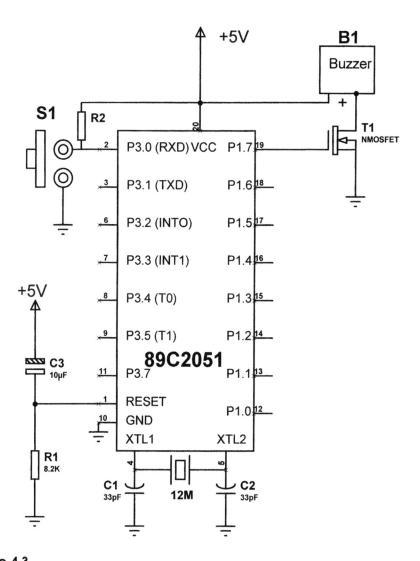

Figure 4.3.

Modified circuit diagram for higher current buzzers

Components Required

In addition to the components used for the basic microcontroller circuit, the following components will be required:

B1	small buzzer (e.g. TDB-05PN)
R2	100K, 0.5 W resistor
T1	VN66AFD MOSFET (optional)
S1	push-button switch

PROJECT 13 – Small Speaker Interface (Using the Timer Interrupt)

Function

This project shows how we can interface our microcontroller to a small speaker type audible device. In this project a continuous single tone output is produced on the speaker when a push-button switch is activated. Timer interrupt of the microcontroller is used to generate the time delay required for the tone. In this project the frequency of the generated tone is 1 kHz (i.e. a period of 1 ms).

Circuit Diagram

The circuit diagram of this project is same as the one in Project 12 (i.e. Fig. 4.3) except that the buzzer is replaced with a small speaker. Bit 7 of port 1 is connected directly to a small speaker via a MOSFET transistor. The port output is in voltage mode so that the speaker will turn on when the port output is at logic HIGH (+5 V). Bit 0 of port 3 is connected to a push-button switch which is normally held at logic HIGH by the pull-up resistor R2.

Program Description

The speaker is initially turned OFF. The push-button switch is then checked and when the switch is pressed, timer 1 of the microcontroller is initialized to generate interrupts at regular intervals. When a timer interrupt is generated the state of the timer is reversed. i.e. if the timer is on, it is turned off and if it is off, it is turned on. The frequency of this waveform is set to be in the audible range and thus it generates an audible sound on the speaker.

The following PDL describes the functions of the program:

Sound Projects

Main program

START
    turn OFF speaker
    IF push-button switch is pressed **THEN**
        Initialize timer 1 to generate interrupts every 250 μs
        Wait for timer interrupt
    ENDIF
END

Timer 1 initialization

START
    Enable timer 1 interrupts
    Set timer 1 to mode 8-bit auto-reload
    Load timer value 6 (i.e. count of 250 μs) into timer register
    Enable microcontroller interrupts
    Turn on timer 1
END

Timer 1 interrupt service routine

START
    IF 500 μs has elapsed **THEN**
        Complement speaker output
    ENDIF
END

Program Listing

The full program listing is shown in Fig. 4.4. When the program starts a variable called *count* is set to 0 and the speaker is turned off. The status of the push-button switch is then checked. If the switch is pressed, function *init_timer* is called to initialize timer 1 of the microcontroller.

Init_timer routine enables timer 1 of the microcontroller, sets timer 1 into 8-bit auto-reload mode (mode 2) and loads the timer counter with 6 so that a timer overflow will occur after 250 counts (i.e. when the timer rolls over from 256 to 0). The timer is then automatically reloaded with the same value. With a microcontroller operating at 12 MHz, the timer clock cycle time is 1 μs since the clock is divided by 12 internally. Thus, a timer interrupt will be generated after every 250 μs. When a timer interrupt is generated, control is directed to the interrupt service routine (ISR) called *timer1( )* as shown in Fig. 4.4. Note that timer 1 interrupt number is 3. The ISR increments the global variable *count*. Variable *count* reaches 2 after two interrupts, i.e. after 500 μs has

/***
PROJECT: PROJECT 13
FILE: PROJ13.C
DATE: August 1999
PROCESSOR: AT892051

This is a simple speaker-based microcontroller project. A miniature speaker is connected to bit 7 of port 1. The speaker normally operates when an alternating signal is applied with the frequency in the audible range. A push-button switch is connected to bit 0 of port 3 and the speaker turns ON when this switch is pressed.

The speaker is connected to bit 7 of port 1 via a MOSFET transistor and the speaker is ON when the output of the port is at logic HIGH.

Timer 1 is used to generate a square wave with a period of 1 ms (i.e. frequency 1 kHz).
***/

```c
#include <AT892051.h>

sbit SPEAKER=P1^7;
sbit PUSH_BUTTON =P3^0;

int count;

/* Timer 1 initialization routine */
void init_timer()
{
    ET1=1;              /*Enable timer 1 int*/
    TMOD=0x20;          /*Timer 1 in Mode 2*/
    TH1=0x6;            /*250 µs count*/
    EA=1;               /*Enable interrupts*/
    TR1=1;              /*Turn on timer 1*/
}

/* Timer 1 interrupt service routine */
timer1() interrupt 3
{
    count++;            /*Inc. count*/
    if(count == 2)      /*count=2*/
    {
        count=0;        /*Reset count*/
        SPEAKER=~SPEAKER;
    }
}

/* Start of main program */
main()
```

```
{
    count=0;                    /*Initialize count*/
    SPEAKER=0;                  /*Speaker OFF*/

    while(PUSH_BUTTON == 1)
    {
    }

    init_timer();               /*Initialize timer*/

    /* Endless loop. Wait here for timer interrupt */
    for(;;)
    {
    }
}
```

Figure 4.4.

Program listing of Project 13

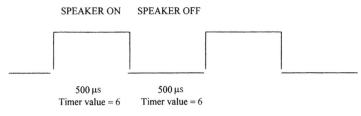

Figure 4.5.

Output waveform produced by Project 13

elapsed. The state of the speaker is then changed after 500 μs. As shown in Fig. 4.5, the period of the generated waveform is thus 1 ms (500 μs ON time and 500 μs OFF time, i.e. a frequency of 1 kHz).

You can change the frequency of the tone easily by loading a different value into the timer register.

Components Required

In addition to the components used for the basic microcontroller circuit, the following components will be required:

B1	small speaker (e.g. T70L015H)
R2	100K, 0.5 W resistor
T1	VN66AFD MOSFET
S1	push-button switch

PROJECT 14 – Two-tone Small Speaker Interface (Using the Timer Interrupt)

Function

This project shows how we can interface our microcontroller to a small speaker type audible device and generate two different tones. When power is applied to the circuit, a continuous single tone of frequency 1 kHz is output to the speaker. When a push-button switch is pressed, the tone frequency is changed to 500 Hz. Timer 1 interrupt of the microcontroller is used to generate the time delay required for the tones.

Circuit Diagram

The circuit diagram of this project is the same as the one in Project 12 (i.e. Fig. 4.3) except that the buzzer is replaced with a small speaker. Bit 7 of port 1 is connected directly to a small speaker via a MOSFET transistor. The port output is in voltage mode so that the buzzer will turn on when the port output is at logic HIGH (+5 V). Bit 0 of port 3 is connected to a push-button switch, which is normally held at logic HIGH by the pull-up resistor R2.

Program Description

The speaker is initially turned OFF. Timer 1 of the microcontroller is then initialized to generate a continuous tone with a frequency of 1 kHz, as in Project 13. The push-button switch is then checked and when the switch is pressed, the timer register value is doubled, i.e. the interrupt interval is increased from 500 µs to 1 ms. A waveform with a 1 ms on and 1 ms off time has a frequency of 500 Hz.

The following PDL describes the functions of the program:

Main program

START
    Turn OFF speaker
    Initialize timer 1 to generate interrupts at 500 µs
    (int_rate = 2)
    DO FOREVER
        IF push-button switch is pressed **THEN**
            Reload timer register for 1 ms interrupts
            (int_rate = 4)
        ENDIF
    ENDDO
END

Timer 1 initialisation

START
      Enable timer 1 interrupts
      Set timer 1 to mode 8-bit auto-reload
      Load timer value 6 (i.e. count of 250 µs) into timer register
      Enable microcontroller interrupts
      Turn on timer 1
END

Timer 1 interrupt service routine

START
      IF (int_rate*250) µs has elapsed **THEN**
            Complement speaker output
      ENDIF
END

Program Listing

The full program listing is shown in Fig. 4.6. When the program starts a variable called *count* is set to 0 and the speaker is turned off. Timer 1 is then initialized with the *int_rate* = 2 so that interrupts are generated every 500 µs, i.e. an output frequency of 1 kHz. The state of the push-button switch is then checked. If the switch is pressed, the interrupt rate, *int_rate*, is changed to 4 so that interrupts will be generated at every 4 × 250 µs = 1 ms i.e. the frequency of the generated waveform is changed to 500 Hz (1 ms on time and 1 ms off time).

timer1( ) is the timer 1 interrupt service routine with interrupt number 3. In this routine, variable *count* is incremented and compared with the *int_rate*. When the two are equal, the speaker output is complemented, i.e. if the speaker is on it is turned off, and if off it is turned on.

The frequency of the generated tones can easily be changed by loading a different value into the timer register or by changing the value of variable *int_rate*.

PROJECT 15 – Electronic Siren (Using the Timer Interrupt)

Function

This project shows how we can interface our microcontroller to a small speaker type audible device and generate a siren sound. When power is

```
/*******************************************************************************
PROJECT:            PROJECT 14
FILE:               PROJ14.C
DATE:               August 1999
PROCESSOR:          AT89205 1

This is a simple speaker-based project. A miniature speaker is connected to bit 7 of port 1.
The speaker normally operates when a varying signal is applied with the frequency in the
audible range. A push-button switch is connected to bit 0 of port 3.

When power is applied to the circuit, a 1 kHz audio signal is sent to the speaker. When the
push-button switch is pressed, the signal frequency is changed to 500 Hz (period = 2 ms).

The speaker is connected to bit 7 of port 1 via a MOSFET transistor and the speaker is ON
when the output of the port is at logic HIGH.

Timer 1 is used to control the period of the square wave signals generated.
*******************************************************************************/
#include <AT892051.h>

sbit SPEAKER=P1^7;
sbit PUSH_BUTTON =P3^0;

int count,int_rate;

/* Timer 1 initialization routine */
void init_timer()
{
   ET1=1;                       /*Enable timer 1 int.*/
   TMOD=0x20;                   /*Timer 1 in Mode 2*/
   TH1=0x6;                     /*Load for 250 µs count*/
   EA=1;                        /*Enable interrupts*/
   TR1=1;                       /*Turn on timer 1*/
}

/* Timer 1 interrupt service routine */
timer1() interrupt 3
{
   count++;                     /*Inc. count*/
   if(count == int_rate)
   {
      count=0;                  /*Reset count*/
      SPEAKER=~SPEAKER;
   }
}

/* Start of main program */
main()
```

```
{
   count=0;                              /*Initialize count*/
   SPEAKER=0;                            /*Speaker OFF*/
   Int_rate=2;                           /*Set for 500 µs*/
   init_timer();                         /*Initialize timer*/

   for(;;)                               /*Endless loop*/
   {
      if(PUSH_BUTTON == 0)int_rate=4;    /*Set to 1 ms*/
   }
}
```

Figure 4.6.

Program listing of Project 14

applied to the circuit, a continuous siren type sound is output from the speaker. Both timer 0 and timer 1 of the microcontroller are used to generate the required tones.

Circuit Diagram

The circuit diagram of this project is the same as the one in Project 12 (i.e. Fig. 4.3) but there is no push-button switch and the buzzer is replaced with a small speaker. Bit 7 of port 1 is connected directly to a small speaker via a MOSFET transistor. The port output is in voltage mode so that the speaker will turn on when the port output is at logic HIGH (+5 V).

Program Description

In this project the frequency of the generated tone is varied continuously from 500 Hz to 10 kHz, thus producing a siren sound. Both timer 0 and timer 1 run at the same time and generate interrupts. Timer 1 generates the output tone and timer 0 changes the frequency of the generated tone continuously. This is how the timers operate:

Timer 1 is in 8-bit auto-reload mode and the timer register is loaded with 50 µs (count of 206). Thus a timer 1 interrupt is generated every 50 µs. Inside the timer 1 interrupt service routine, a counter is incremented and its value compared to a global variable called *int_rate*. When the two are equal the speaker output is changed. The frequency of the generated output waveform is then as follows (notice that the period is twice the timer count value since half of the period is off and the other half is on):

int_rate = 1	period = 100 μs	frequency = 10 kHz
int_rate = 2	period = 200 μs	frequency = 5 kHz
int_rate = 4	period = 400 μs	frequency = 2.5 kHz
int_rate = 5	period = 500 μs	frequency = 2 kHz

In general, we can say that the frequency of the generated tone is given by:

freq = 10/int_rate

where freq is in kHz.

In this project, variable *int_rate* is varied from 100 down to 1, i.e. the frequency of the generated tone varies between 100 Hz and 10 kHz.

Timer 0 of the microcontroller is used to change the frequency of the tone by changing the value of variable *int_rate*. Timer 0 is in 8-bit auto-reload mode and the timer register is loaded with 56 so that it generates interrupts at every 200 μs. Inside the timer 0 interrupt service routine a counter is used and variable *int_rate* is decremented by 1 when the counter counts by 200. Thus, variable *int_rate* will be decremented every 200 × 200 μs = 40,000 μs or 40 ms.

In summary, the frequency of the generated tone will vary every 20 ms from 100 Hz to 10 kHz. The result is that a siren type output will be generated on the speaker.

The following PDL describes the functions of the program:

Main program

START
    Turn OFF speaker
    Initialize timer 1 for auto-reload 50 μs interrupts
    Initialize timer 0 for auto-reload 200 μs interrupts
    Set *int_rate* for 100 Hz
    DO FOREVER
        Wait for timer interrupts
    ENDDO

Timer initialization routine

START
    Enable timer 1 interrupts
    Set timer 1 to mode 8-bit auto-reload
    Load timer 1 with 206 (i.e. count of 50 μs)

Enable timer 0 interrupts
Set timer 0 to mode 8-bit auto-reload
Load timer 0 with 56 (i.e. count of 200 μs)
Enable timer 0 interrupts
Turn timer 1 on
Turn timer 0 on
END

Timer 1 interrupt service routine

START
    IF *int_rate* μs has elapsed **THEN**
        Complement speaker output
    ENDIF
END

Timer 0 interrupt service routine

START
    IF 40 ms has elapsed **THEN**
        Decrement *int_rate*
        IF *int_rate* = 0 **THEN**
            Set *int_rate* for 100 Hz
        ENDIF
    ENDIF
END

Program Listing

The full program listing is shown in Fig. 4.7. Variable *SPEAKER* is assigned to bit 7 of port 1. When the program starts, variables *count* and *timer1_overflow* are set to 0. The speaker is then turned off. Timers 0 and 1 are initialized by calling function *init_timers*. The program then enters an endless loop and waits until the timer interrupts occur.

Inside the timer 1 interrupt service routine, variable count is incremented and compared to variable *int_rate*. When the two are equal, count is reset to 0 and the speaker output is complemented.

Inside the timer 0 interrupt service routine, variable *timer1_overflow* is incremented and when it reaches 200 (i.e. 200 × 200 μs = 40 ms), it is reset to 0 and *int_rate* is decremented so that a higher frequency tone could be generated by timer 1. When *int_rate* reaches 0 it is reset back to 100 so that the process can repeat.

100 Microcontroller Projects in C for the 8051

```
/****************************************************************************
PROJECT:                PROJECT 15
FILE:                   PROJ15.C
DATE:                   August 1999
PROCESSOR:              AT892051
```

This is a simple speaker-based siren project. A miniature speaker is connected to bit 7 of port 1. The speaker normally operates when an alternating signal is applied with the frequency in the audible range.

When power is applied to the circuit, a 100 Hz audio signal is first sent to the speaker using timer 1. Timer 0 then runs and changes the frequency of the tone from 100 Hz to 10 kHz, every 40 ms. The effect is that an audible siren type output is produced.

The speaker is connected to bit 7 of port 1 via a MOSFET transistor and the buzzer is ON when the output of the port is at logic HIGH.

Timers 0 and 1 are used to generate variable frequency and variable pitch output.
```
****************************************************************************/
#include <AT892051.h>

sbit SPEAKER=P1^7;

int count,timer1_overflow,int_rate;

/* Timer initialization routine */
void init_timers()
{
   ET1=1;                   /*Enable timer 1 interrupt*/
   TMOD=0x20;               /*Timer 1 in Mode 2*/
   TH1=206;                 /*Timer 1 50 µs count*/
   ET0=1;                   /*Enable timer 0 int.*/
   TMOD=TMOD | 2;           /*Timer 0 in Mode 2*/
   TH0=56;                  /*Timer 0 200 µs count*/
   EA=1;                    /*Enable interrupts*/
   TR1=1;                   /*Turn on timer 1*/
   TR0=1;                   /*Turn on timer 0*/
}

/* Timer 1 interrupt service routine */
timer1() interrupt 3
{
   count++;                 /*Inc. count*/
   if(count == int_rate)
   {
      count=0;              /*Reset count*/
      SPEAKER=~SPEAKER;
   }
}
```

```
/* Timer 0 interrupt service routine */
timer0() interrupt 1
{
  timer1_overflow++;
  if(timer1_overflow == 200)    /*if 40 ms */
  {
    timer1_overflow=0;
    int_rate--;
    count=0;                    /*clear count*/
    if(int_rate == 0)int_rate=100;
  }
}

/* Start of main program */
main()
{
  count=0;                      /*Initialize count*/
  timer1_overflow=0;
  SPEAKER=0;                    /*Speaker OFF*/
  int_rate=100;                 /*Set for 100 Hz */
  init_timers();                /*Initialize timers*/

    for(;;)                     /*Endless loop*/
    {
    }
}
```

Figure 4.7.
Program listing of Project 15

PROJECT 16 – Electronic Organ (Using the Timer Interrupt)

Function

This is a simple electronic organ project. A small speaker is connected to bit 0 of port 3 via a MOSFET transistor. Eight push-button switches are connected to port 1 to act as the keyboard for the electronic organ. Timer 1 of the microcontroller is used to generate time delays for the required frequencies. Only one octave (eight notes) is provided.

Circuit Diagram

The circuit diagram of this project is shown in Fig. 4.8. The speaker is connected to bit 0 of port 3. The keyboard switches are connected to port 1.

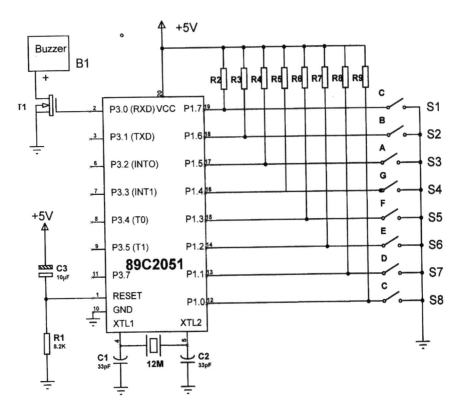

Figure 4.8.
Circuit diagram of Project 16

Bit 0 is assigned to note C, bit 1 is assigned to note D, bit 2 is assigned to note E and so on. The switches are normally held at logic HIGH with pull-up resistors (R2 to R9). Pressing a switch sends a logic LOW to the microcontroller port.

Program Description

In this project timer 1 is used to generate the tones for the electronic organ. The timer is in auto-reload mode and loaded to generate an interrupt every 50 µs (i.e. loaded with 206 for a 12 MHz crystal). An endless *for* loop is formed and the keyboard is scanned. When a key is pressed, the timer is reloaded with the correct value so that the required tone can be generated.

The following octave was used for this project (the frequencies are in Hz):

Notes: C   D   E   F   G   A   B   C
Freq: 262 294 330 349 392 440 494 524

The periods of the notes in μs are then given by (period = 1/frequency):

Notes: C   D   E   F   G   A   B   C
Period: 3816 3401 3030 2865 2551 2272 2024 1908

The counter is loaded so that it generates an interrupt every 50 μs. The number of counts (N) required for each note is therefore given by dividing the period by 50 μs, as shown below:

Notes: C   D   E   F   G   A   B   C
N: 76 68 60 57 51 45 40 38

The following PDL describes the functions of the program:

Main program

START
    Turn off speaker
    Initialize timer 1
    Initialize *count*
    DO FOREVER
        IF a key is pressed **THEN**
            Load value of the key into variable *tone*
        ENDIF
    ENDDO
END

Timer 1 interrupt service routine

START
    Increment *count*
    IF *count* = *tone* **THEN**
        count = 0
        Complement speaker output
    ENDIF
END

Program Listing

The full program listing is shown in Fig. 4.9. When the program is started, the speaker is turned off and timer 1 is initialized to 8-bit auto-reload mode by calling function *init_timer*. An endless *for* loop is then formed and the keyboard is scanned. Normally port 1 contains the value 0xFF (255) when no keys are pressed. When a key is pressed the pin corresponding to that key goes to logic LOW. Variable *key_pressed* reads port 1 and complements the

```
/********************************************************************
PROJECT:                PROJECT 16
FILE:                   PROJ16.C
DATE:                   August 1999
PROCESSOR:              AT892051

This is a simple electronic organ project. A small speaker is connected to bit 0 of port 3 via
a MOSFET transistor. The speaker normally operates when an alternating signal is applied
with the frequency in the audible range.

Eight push-button switches are connected to port 1 to act as the keyboard for the
electronic organ. Timer 1 is used to generate time delays for the required frequencies.
********************************************************************/
#include <AT892051.h>

sbit SPEAKER=P3^0;

int count,tone;

/* Timer 1 initialization routine */
void init_timer()
{
  ET1=1;                    /*Enable timer 1 int.*/
  TMOD=0x20;                /*Timer 1 in Mode 2*/
  TH1=206;                  /*Timer 1 50 µs count*/
  TR1=1;                    /*Run timer 1*/
  EA=1;                     /*Enable interrupts*/
}

/* Timer 1 interrupt service routine */
timer1() interrupt 3
{
  count++;                  /*Increment count*/
  if(count == tone)
  {
    count=0;                /*Reset count*/
    SPEAKER=~SPEAKER;
  }
}

/* Start of main program */
main()
{
  unsigned char key_pressed;
  count=0;                  /*Initialize count*/
  SPEAKER=0;
  init_timer();             /*Initialize timers*/

  for(;;)                   /*Endless loop*/
```

```
{
  if(P1 != 0xFF)              /*If a key pressed*/
  {
    TR1=1;                    /*Turn on timer*/
    key_pressed=~P1;          /*Complement key*/

    /* Check which key pressed */
    switch(key_pressed)
    {
    case 1:                   /*If key 1 is pressed...*/
          tone=76;
          break;
    case 2:
          tone=68;
          break;
    case 4:
          tone=60;
          break;
    case 8:
          tone=57;
          break;
    case 16:
          tone=51;
          break;
    case 32:
          tone=45;
          break;
    case 64:
          tone=40;
          break;
    case 128:
          tone=38;
          break;
    }
  }
  else
  {
    /* No key pressed */
    SPEAKER=0;
    count=0;                  /*Reset count*/
    TR1=0;                    /*Stop timer*/
  }
 }
}
```

Figure 4.9.

Program listing of Project 16

value read so that the numbers obtained correspond to the key numbers as powers of 2. Thus, as an example, when key C is pressed, *key_pressed* contains 1, when D is pressed, *key_pressed* is 2, when key E is pressed, *key_pressed* is 4 and so on. A *switch* statement is used to load variable *tone* with the correct timer value so that the required tone can be generated on the speaker.

Components Required

In addition to the standard components used for the microcontroller, the following components will be required for this project:

S1 to S8   SPDT switches
R2 to R9   100K, 0.125 W resistors
B1   small speaker
T1   n-channel MOSFET transistor (e.g. VN66AFD)

CHAPTER 5

TEMPERATURE PROJECTS

Temperature measurement and control is one of the most common applications of microcontroller-based data acquisition systems. Four types of sensors are commonly used to measure temperature in commercial and industrial applications. These are *thermocouples, resistive temperature devices* (RTDs), *thermistors*, and *integrated circuit* (IC) *temperature sensors*. Each sensor has its unique advantages and disadvantages and by understanding how these sensors work, and what types of signal conditioning are required for each, we can make more accurate and reliable temperature measurement, monitoring, and control.

The typical characteristics of various temperature sensors are:

- *Thermocouples*: these are inexpensive, and the most common temperature sensors with a wide range of temperature range. Thermocouples work on the principle that when two dissimilar metals are combined, a voltage appears across the junction between the metals. By measuring this voltage, we can get a temperature reading. Different combinations of metals create different thermocouple voltages and there is a wide range of thermocouples available for different applications. Thermocouples generate very low voltages, typically 50 μV/°C. These low-level signals require special signal conditioning to remove any possible noise. Thermocouples have non-linear relationships to the measured temperature and as a result it is necessary either to linearize the characteristics or to use look-up tables to obtain the actual temperature from the measured voltage.

- *RTDs*: an RTD is a resistor with its resistance changing with temperature. The most popular type of RTD is made of platinum and has a resistance of 100 Ω at 0°C. Because RTDs are resistive devices, a current must pass through the RTD to produce a voltage that can be measured. The change in resistance is very small (about 0.4 Ω/°C) and special circuitry is generally needed to measure the small changes in temperature. One of the drawbacks of RTDs is their non-linear change in resistance with temperature.

- *Thermistors*: thermistors are metal oxide semiconductor devices whose resistance changes with temperature. One of the advantages of thermistors is their fast responses and high sensitivity. For example, a typical thermistor may have a resistance of 50 kΩ at 25°C, but have a resistance of only 2 kΩ at 85°C. Like RTDs, a current is passed through a thermistor and the voltage across the thermistor is measured. Thermistors are very non-linear devices and look-up tables are usually used to convert the measured voltage to temperature. Thermistors are very small and one disadvantage of this is that they can be self-heating under a large excitation current. This of course increases the temperature of the device and can give erroneous results.

- *IC temperature sensors*: integrated circuit temperature sensors are usually 3- or 8-pin active devices that require a power supply to operate and give out a voltage which is directly proportional to the temperature. There are basically two types of IC temperature sensor: analogue sensors are usually 3-pin devices and give out an analogue voltage of typically 10 mV/°C which is directly proportional to the temperature; digital temperature sensors provide 8- or 9-bit serial digital output data which is directly proportional to the temperature.

In this chapter we shall be looking at how we can interface various temperature sensors to our microcontroller in order to measure and display the ambient temperature.

PROJECT 17 – Using a Digital Temperature Sensor

Function

This project shows how we can interface a DS1620 type digital temperature sensor to our microcontroller. The ambient temperature is measured continuously and then displayed in degrees centigrade on three TIL311 type alphanumeric displays. Positive temperature is displayed from 0°C to 125°C. Negative temperature is displayed with a leading letter 'E' in the range down to −55°C.

Circuit Diagram

The block diagram of this project is shown in Fig. 5.1. DS1620 is a digital IC temperature sensor which measures the ambient temperature and provides the output as 9 bits of digital serial data. The microcontroller extracts the temperature data from the DS1620 and then displays the temperature on three TIL311 type alphanumeric displays.

Before describing the circuit diagram in detail, it is useful to look at the operation of the DS1620 temperature sensor IC.

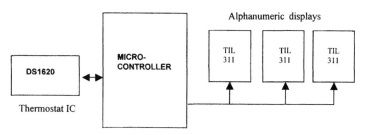

Figure 5.1.
Block diagram of Project 17

DS1620 Digital Thermometer IC

DS1620 is a digital thermometer and thermostat IC that provides 9 bits of serial data to indicate the temperature of the device. The pin configuration of the DS1620 is shown in Fig. 5.2. VDD is the power supply which is normally connected to a +5 V supply. DQ is the data input/output pin. CLK is the clock input. RST is the reset input. The device can also act as a thermostat. THIGH is driven high if the DS1620's temperature is greater than or equal to a user defined temperature TH. Similarly, TLOW is driven high if the DS1620's temperature is less than or equal to a user defined temperature TL. TCOM is driven high when the temperature exceeds TH and stays high until the temperature falls below TL. User defined temperatures TL and TH are stored in non-volatile memory of the device so that they are not lost even after removal of the power.

Data is output from the device as 9 bits, with the LSB sent out first. The temperature is provided in 2's complement format from −55°C to +125°C, in steps of 0.5°C. Table 5.1 shows the relationship between the temperature and data output by the device.

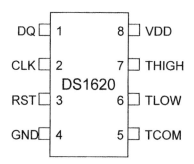

Figure 5.2.
Pin configuration of DS1620

Table 5.1 Temperature/data relationship of DS1620

Temp. (°C)	Digital output (binary)	Digital output (hex)	2's complement	Digital output (decimal)
+125	0 11111010	0FA	–	250
+25	0 00110010	032	–	50
0.5	0 00000001	001	–	1
0	0 00000000	000	–	0
−0.5	1 11111111	1FF	001	511
−25	1 11001110	1CE	032	462
−55	1 10010010	192	06E	402

Operation of DS1620

Data input and output is through the DQ pin. When RST input is high, serial data can be written or read by pulsing the clock input. Data is written or read from the device in two parts. First, a protocol is sent and then the required data is read or written. The protocol is 8-bit data and the protocol definitions are given in Table 5.2. For example, to write the thermostat value TH, the hexadecimal protocol data 01 is first sent to the device. After issuing this

Table 5.2 DS1620 Protocol definitions

PROTOCOL	PROTOCOL DATA (hex)
Write TH	01
Write TL	02
Write configuration	0C
Stop conversion	22
Read TH	A1
Read TL	A2
Read temperature	AA
Read configuration	AC
Start conversion	EE

command, the next nine clock cycles clock in the 9-bit temperature limit which will set the threshold for operation of the THIGH output.

For example, the following data (in hexadecimal) should be sent to the device to set it for a TH limit of +50°C and TL limit of +20°C and then subsequently to start the conversion:

01   Send TH protocol
64   Send TH limit of 50 (64 hex = 100 decimal)
02   Send TL protocol
28   Send TL limit of 20 (28 hex = 40 decimal)
EE   Send conversion start protocol

A configuration/status register is used to program various operating modes of the device. This register is written with protocol 0C (hex) and the status is read with protocol AC (hex). Some of the important configuration/status register bits are as follows:

Bit 0:   This is the 1 shot mode. If this bit is set, the DS1620 will perform one temperature conversion when the start convert protocol is sent. If this bit is 0, the device will perform continuous temperature conversions.
Bit 1:   This bit should be set to 1 for operation with a microcontroller or microprocessor.
Bit 5:   This is the TLF flag and is set to 1 when the temperature is less than or equal to the value TL.
Bit 6:   This is the THF bit and is set to 1 when the temperature is greater than or equal to the value of TH.
Bit 7:   This is the DONE bit and is set to 1 when a conversion is complete.

The complete circuit diagram of this project is shown in Fig. 5.3. Bit 2 of port 3 is connected to the RST input of DS1620, bit 1 is connected to the clock input and bit 0 of port 3 is connected to the DQ pin of the DS1620. Three TL311 type alphanumeric displays are connected to port 1 of the microcontroller. Digit 1 is controlled from bit 7 of port 1, digit 2 from bit 6 of port 1, and digit 3 from bit 5 of port 1.

Program Description

The program reads the temperature from the DS1620 thermometer IC and displays the temperature on three TIL311 type displays continuously with 1 second delay between each displayed output. The following PDL describes the operation of the program:

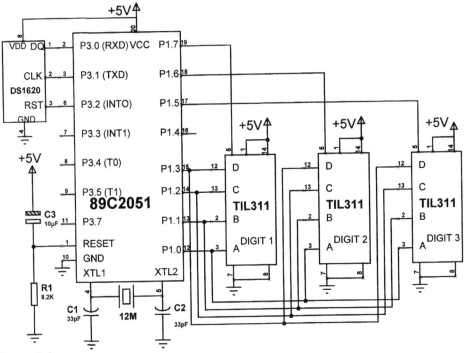

Figure 5.3.
Circuit diagram of Project 17

Main program

    START
        Set digit latches to 1
        Configure DS1620
        Start temperature conversion
        DO FOREVER
            Read temperature
            Display temperature
            Wait a second
        ENDDO
    END

Function configure DS1620

    START
        Set configuration/status register to 2 (i.e. continuous operation)
    END

Function start temperature conversion

START
    Send protocol EE (hex) to start temperature conversion
END

Function read temperature

START
    Call function read_from_ds1620 to get the temperature
END

Function read_from_ds1620

START
    Set RST bit to 1
    Read 9-bit serial temperature data from DS1620
    Set RST bit to 0
END

Function display temperature

START
    IF temperature is negative **THEN**
        Get 2's complement of the temperature reading
        Divide temperature by 2 to get real temperature
        Set digit 1 to display letter 'E'
        Display temperature digits
    ELSE
        Divide temperature by 2 to get real temperature
        Display temperature digits
    ENDIF
END

Program Listing

The full program listing is given in Fig. 5.4. The display digit latches and the DS1620 control lines are assigned to bit variables at the beginning of the program. Also the used protocols are defined and assigned to global variables. For example, *read_temp* is assigned to hexadecimal number AA, *start_conv* is assigned to hexadecimal number EE and so on.

When the program starts, the digit latches *digit1_latch*, *digit2_latch*, and *digit3_latch* are all set to 1 to avoid any erroneous writes to the displays. A function *configure_ds1620* is then called to set the configuration register/status

114 Microcontroller Projects in C for the 8051

```
/****************************************************************************
PROJECT:                PROJECT 17
FILE:                   PROJ17.C
DATE:                   August 1999
PROCESSOR:              AT892051

This is a temperature monitoring project. A DS1620 type digital thermometer is used to
read the ambient temperature. The temperature is then displayed on three TL311 type
alphanumeric displays. The temperature range is -55°C to +125°C. Positive temperature
is displayed with leading zeros. Negative temperatures are displayed by inserting the
letter 'E' in front of the display. The display accuracy is +/- 1°C, i.e. there is no decimal
point in the displayed data.

The display is updated every second.
****************************************************************************/
#include <AT892051.h>

sbit digit1_latch = P1^7;           /*digit 1 latch*/
sbit digit2_latch = P1^6;           /*digit 2 latch*/
sbit digit3_latch = P1^5;           /*digit 3 latch*/

sbit ds1620_dq = P3^0;              /*DS1620 data pin*/
sbit ds1620_clk = P3^1;             /*DS1620 clock pin*/
sbit ds1620_rst = P3^2;             /*DS1620 reset pin*/

#define read_temp 0xAA              /*read temp command*/
#define start_conv 0xEE             /*start conversion command*/
#define write_config 0x0C           /*write config command*/

/* Function to delay about a second */
void wait_a_second()
{
    unsigned int x;
    for(x=0;x<33000;x++);
}

/* Function to display data on three TIL311 displays. Negative temperature is displayed
with a leading 'E' character. Display range is -55 to +125. */
void display_temperature(unsigned int x)
{
    unsigned int s;
    int first,second,third;
    if(x > 255)                     /*if negative*/
    {
        x=~x;                       /*complement temp*/
        x++;                        /*add 1 for 2s comp*/
        x=x & 0xFF;                 /*extract lower 8 bits*/
        x=x/2;                      /*get real temp*/
```

```
    first=14;                   /*display leading 'E'*/
    second=x/10;
    third=x-10*second;
  }
  else
  {
    x=x/2;                      /*temp is positive*/
    first=x/100;                /*extract digit data*/
    s=x-100*first;
    second=s/10;
    third=s-second*10;
  }

  P1=first | 0xE0;              /*Send digit1 data*/
  digit1_latch=0;               /*Latch the digit1 data*/
  digit1_latch=1;               /*Set digit1 latch on*/

  P1=second | 0xE0;             /*Send digit2 data*/
  digit2_latch=0;               /*Latch the digit2 data*/
  digit2_latch=1;               /*Set digit2 latch on*/

  P1=third | 0xE0;              /*Send digit3 data*/
  digit3_latch=0;               /*Latch digit3 data*/
  digit3_latch=1;               /*Set digit3 latch on*/
}

/* This function sends a data bit to DS1620 thermometer IC */
void write_ds1620_bit(unsigned char b)
{
  ds1620_dq=b;                  /*send bit*/
  ds1620_clk=0;                 /*set clock 0*/
  ds1620_clk=1;                 /*set clock 1*/
  ds1620_dq=1;                  /*set data 1*/
}

/* This function reads a data bit from DS1620 */
unsigned char read_ds1620_bit()
{
  unsigned char b;

  ds1620_clk=0;                 /*set clock 0*/
  b=ds1620_dq;                  /*read a bit*/
  ds1620_clk=1;                 /*set clock 1*/
  return (b);                   /*return bit*/
}

/* This function writes data/configuration to DS1620 */
```

```c
void write_to_ds1620(unsigned char ds1620_function,
   unsigned char ds1620_data,
   unsigned char bit_count)
{
   unsigned char i,this_bit;

   ds1620_rst=1;                      /*set reset to 1*/
   for(i=0;i<8;i++)                   /*send function...*/
   {
      this_bit=ds1620_function >> i;
      this_bit=this_bit & 0x01;
      write_ds1620_bit(this_bit);
   }
   for(i=0;i<bit_count;i++)           /*send data...*/
   {
      this_bit=ds1620_data >> i;
      this_bit=this_bit & 0x01;
      write_ds1620_bit(this_bit);
   }
   ds1620_rst=0;
}                                     /*set reset to 0*/

/* This function reads data/configuration from the DS1620 */
unsigned int read_from_ds1620(unsigned char ds1620_function,
   unsigned char bit_count)
{
   unsigned char i,this_bit;
   unsigned int ds1620_data;

   ds1620_data=0;                     /*initialize data*/
   ds1620_rst=1;                      /*set reset to 1*/
   for(i=0;i<8;i++)
   {                                  /*send function...*/
      this_bit=ds1620_function >> i;
      this_bit=this_bit & 0x01;
      write_ds1620_bit(this_bit);
   }                                  /*read data*/
   for(i=0;i<bit_count;i++)
   {
      ds1620_data=ds1620_data | read_ds1620_bit() << i;
   }
   ds1620_rst=0;
   return (ds1620_data);
}

/* This function configures the DS1620 for continuous operation */
void configure_ds1620()
```

```
{
    write_to_ds1620(write_config,2,8);
    wait_a_second();
}

/* This function starts the conversion */
void start_temp_conv()
{
    write_to_ds1620(start_conv,0,0);
}

/* This function reads the temperature */
unsigned int read_temperature()
{
    unsigned int t;

    t=read_from_ds1620(read_temp,9);    /*read temp*/
    return (t);                          /*return temp*/
}

/* Start of main program */
main()
{
    unsigned int TEMP;
    digit1_latch=1;           /*Set digit1 latch*/
    digit2_latch=1;           /*set digit2 latch*/
    digit3_latch=1;           /*set digit3 latch*/

    configure_ds1620();       /*configure DS1620*/
    start_temp_conv();        /*start conversion*/

    for(;;)                   /*endless loop*/
    {
        TEMP=read_temperature();      /*read temperature*/
        display_temperature(TEMP);    /*Output to TIL311*/
        wait_a_second();              /*wait a second*/
    }
}
```

Figure 5.4.

Program listing of Project 17

for continuous operation. Temperature conversion is then started by calling the function *start_temp_conversion*. This function sends protocol EE (hex) to the DS1620. An endless loop is then formed using the *for* statement with no parameters. Inside this loop, function *read_temperature* reads the 9-bit temperature value and returns in variable *TEMP*. Function *display_temperature* displays the temperature on the three TIL311 displays. This loop is repeated with about a 1 second delay between each output.

Function *read_temperature* returns the temperature to the calling program as an unsigned integer. This function calls function *read_from_ds1620* with the argument AA (in hex) to get the temperature. Function *read_from_ds1620* is a general routine which reads data from the DS1620. This function sends a protocol data to the DS1620 and then reads data bytes from the DS1620 corresponding to the sent protocol. The *RST* input of the device is first set to 1. A *for* loop is then formed to iterate eight times to send serial protocol data to the DS1620. LSB is sent out first. Local variable *this_bit* stores the bit to be sent out at each iteration. Another *for* loop reads data from the DS1620 and stores this data in variable *ds1620_data*. At the end of the read cycle the *RST* input is set back to 0 and the data in *ds1620_data* is returned to the caller.

Function *write_to_ds1620* is a general function and it sends a protocol, followed by data bits, to the DS1620. The *RST* input of the DS1620 is first set to 1. A *for* loop is then formed to iterate eight times and the protocol bits are sent out serially to the *DQ* input of the DS1620. Variable *this_bit* stores the bit to be sent out at each iteration. After this, another *for* loop sends out the required number of data bits to the DS1620. At the end of the write cycle, the *RST* input is returned to 0.

Function *display_temperature* receives the temperature data as its argument and displays the temperature on the three TIL311 type alphanumeric displays. If the temperature is negative, the first display digit is set to display letter 'E'. The temperature value to be displayed is divided by 2 since the temperature is returned by the DS1620 in 0.5°C intervals. For example, a reading of decimal 100 corresponds to 50°C. Each digit is displayed after converting the data to decimal format.

Components Required

In addition to the basic components required by the microcontroller, the following components will be required for this project:

DS1620   thermometer IC
TIL311   alphanumeric displays (3 off)

PROJECT 18 – Digital Thermometer with Centigrade/Fahrenheit Output

Function

This project is similar to Project 17, but in addition the outputs can display the temperature in both Centigrade (°C) and Fahrenheit (°F). An SPDT switch is connected to bit 3 of port 3 and the output of this switch is normally held at logic high with a pull-up resistor. In this state the output display is in °C. When the switch is pressed, the display changes to show the temperature in °F. The rest of the project is the same as Project 17, i.e. the temperature is measured with a DS1620 type thermometer IC and the output is displayed on three TIL311 type alphanumeric displays.

Circuit Diagram

The block diagram of this project is shown in Fig. 5.5. The circuit diagram is similar to the circuit of Project 17 with the addition of an SPDT switch to bit 3 of port 3. The temperature is sensed by the DS1620 thermometer IC and the output is displayed either in °C or in °F based upon the state of the SPDT switch. The full circuit diagram is shown in Fig. 5.6.

Program Description

The program is the same as the one in Project 17 except that the state of the SPDT switch is monitored and when the switch is pressed, the temperature is converted from °C to °F and then displayed accordingly. Negative temperatures are displayed by inserting the leading letter 'E'.

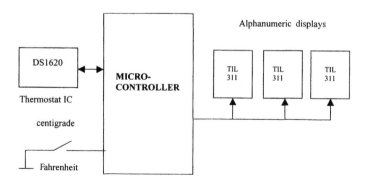

Figure 5.5.

Block diagram of Project 18

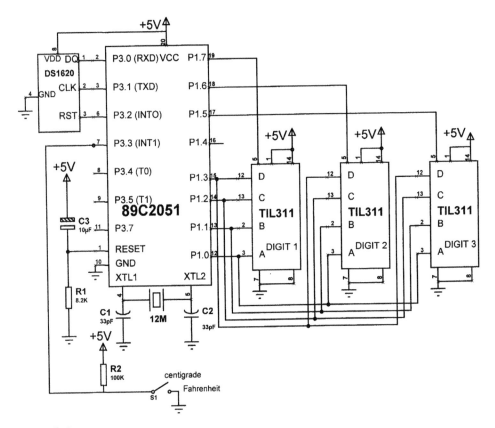

Figure 5.6.
Circuit diagram of Project 18

Program Listing

The full program listing of this project is shown in Fig. 5.7. Only the parts which are different from Project 17 are described here. Variable *mode* is assigned to bit 3 of port 3. The value of *mode* is checked in function *display_temperature*. If *mode* is zero (i.e. in Fahrenheit mode), the temperature is converted to °F by multiplying by 1.8 and adding 32. The final temperature is then displayed as in Project 17.

Components Required

In addition to the standard components required by the microcontroller, the following components will be required for this project:

Temperature Projects

```
/*******************************************************************************
PROJECT:            PROJECT 18
FILE:               PROJ18.C
DATE:               August 1999
PROCESSOR:          AT892051
```

This is a temperature monitoring project. A DS1620 type digital thermometer is used to read the ambient temperature. The temperature is then displayed on three TL311 type alphanumeric displays. The temperature range is −55°C to +125°C. Positive temperature is displayed with leading zeros. Negative temperatures are displayed by inserting the letter 'E' in front of the display. The display accuracy is +/−1°C, i.e. there is no decimal point in the displayed data.

The display is updated every second.

Bit 3 of port 3 is connected to a switch. This switch is normally held at logic 1 with a pull-up resistor. When the switch is 1, the temperature is displayed in degrees centigrade. When the switch is 0, the temperature is displayed in degrees Fahrenheit.
```
*******************************************************************************/
#include <AT892051.h>

sbit digit1_latch = P1^7;          /*digit 1 latch*/
sbit digit2_latch = P1^6;          /*digit 2 latch*/
sbit digit3_latch = P1^5;          /*digit 3 latch*/

sbit mode = P3^3;                  /*centigrade/fahrenheit select*/
sbit ds1620_dq = P3^0;             /*DS1620 data pin*/
sbit ds1620_clk = P3^1;            /*DS1620 clock pin*/
sbit ds1620_rst = P3^2;            /*DS1620 reset pin*/

#define read_temp 0xAA             /*read temp command*/
#define start_conv 0xEE            /*start conversion command*/
#define write_config 0x0C          /*write config command*/

/* Function to delay about a second */
void wait_a_second()
{
    unsigned int x;
    for(x=0;x<33000;x++);
}

/* Function to display data on three TIL311 displays. Negative temperature is displayed
with a leading 'E' character. Display range is −55 to +125. */
void display_temperature(unsigned int x)
{
    unsigned int s;

    int first,second,third;
```

```c
    if(x > 255)                 /*if negative*/
    {
       x=~x;                    /*complement temp*/
       x++;                     /*add 1 for 2s comp*/
       x=x & 0xFF;              /*extract lower 8 bits*/
       x=x/2;                   /*get real temp*/
       if(mode == 0)x=1.8*x+32; /*in fahrenheit*/
       first=14;                /*display leading 'E'*/
       second=x/10;
       third=x-10*second;
    }
    else
    {
       x=x/2;                   /*temp is positive*/
       if(mode == 0)x=1.8*x+32; /*in fahrenheit*/
       first=x/100;             /*extract digit data*/
       s=x-100*first;
       second=s/10;
       third=s-second*10;
    }

    P1=first | 0xE0;            /*Send digit1 data*/
    digit1_latch=0;             /*Latch the digit1 data*/
    digit1_latch=1;             /*Set digit1 latch on*/

    P1=second | 0xE0;           /*Send digit2 data*/
    digit2_latch=0;             /*Latch the digit2 data*/
    digit2_latch=1;             /*Set digit2 latch on*/

    P1=third | 0xE0;            /*Send digit3 data*/
    digit3_latch=0;             /*Latch digit3 data*/
    digit3_latch=1;             /*Set digit3 latch on*/
}

/* This function sends a data bit to DS1620 thermometer IC */
void write_ds1620_bit(unsigned char b)
{
    ds1620_dq=b;                /*send bit*/
    ds1620_clk=0;               /*set clock 0*/
    ds1620_clk=1;               /*set clock 1*/
    ds1620_dq=1;                /*set data 1*/
}

/* This function reads a data bit from DS1620 */
unsigned char read_ds1620_bit()
{
```

```c
   unsigned char b;

   ds1620_clk=0;              /*set clock 0*/
   b=ds1620_dq;               /*read a bit*/
   ds1620_clk=1;              /*set clock 1*/
   return (b);                /*return bit*/
}

/* This function writes data/configuration to DS1620 */
void write_to_ds1620(unsigned char ds1620_function,
                     unsigned char ds1620_data, unsigned char bit_count)
{
   unsigned char i,this_bit;

   ds1620_rst=1;              /*set reset to 1*/
   for(i=0;i<8;i++)           /*send function...*/
   {
      this_bit=ds1620_function >> i;
      this_bit=this_bit & 0x01;
      write_ds1620_bit(this_bit);
   }
   for(i=0;i<bit_count;i++)   /*send data...*/
   {
      this_bit=ds1620_data >> i;
      this_bit=this_bit & 0x01;
      write_ds1620_bit(this_bit);
   }
   ds1620_rst=0;
}                             /*set reset to 0*/

/* This function reads data/configuration from the DS1620 */
unsigned int read_from_ds1620(unsigned char ds1620_function,
                              unsigned char bit_count)
{
   unsigned char i,this_bit;
   unsigned int ds1620_data;

   ds1620_data=0;             /*initialize data*/
   ds1620_rst=1;              /*set reset to 1*/
   for(i=0;i<8;i++)
   {                          /*send function...*/
      this_bit=ds1620_function >> i;
      this_bit=this_bit & 0x01;
      write_ds1620_bit(this_bit);
   }                          /*read data*/
   for(i=0;i<bit_count;i++)
```

```c
    {
       ds1620_data=ds1620_data | read_ds1620_bit() << i;
    }
    ds1620_rst=0;
    return (ds1620_data);
}

/* This function configures the DS1620 for continuous operation */
void configure_ds1620()
{
   write_to_ds1620(write_config,2,8);
   wait_a_second();
}

/* This function starts the conversion */
void start_temp_conv()
{
   write_to_ds1620(start_conv,0,0);
}

/* This function reads the temperature */
unsigned int read_temperature()
{
   unsigned int t;

   t=read_from_ds1620(read_temp,9);    /*read temp*/
   return (t);                          /*return temp*/
}

/* Start of main program */
main()
{
   unsigned int TEMP;

   digit1_latch=1;                      /*Set digit1 latch*/
   digit2_latch=1;                      /*set digit2 latch*/
   digit3_latch=1;                      /*set digit3 latch*/

   configure_ds1620();                  /*configure DS1620*/
   start_temp_conv();                   /*start conversion*/

   for(;;)                              /*Start of endless loop*/

   {
      TEMP=read_temperature();          /*read temperature*/
      display_temperature(TEMP);        /*Output to TIL311*/
      wait_a_second();                  /*wait a second*/
   }
}
```

Figure 5.7.
Program listing of Project 18

DS1620   thermometer IC
TIL311   alphanumeric displays (3 off)
S1   SPDT switch
R2   100K, 0.125 W resistor

PROJECT 19 – Digital Thermometer with High Alarm Output

Function

This project is similar to Project 17 but a buzzer is connected to the THIGH output of the DS1620 thermometer IC, via a MOSFET transistor. When the temperature exceeds a preset value the buzzer turns on and stays on as long as the temperature is above this value. In this project the alarm sounds when the temperature is equal to or greater than 25°C.

Circuit Diagram

The circuit diagram of this project is shown in Fig. 5.8. The DS1620 thermometer IC and the TIL311 displays are connected as in Projects 17 and 18. A small buzzer is connected to the THIGH output of the DS1620 via a MOSFET power transistor. Normally the THIGH output is at logic low level and this output goes to logic high when the temperature exceeds the value TH stored in the non-volatile memory of the DS1620.

Program Description

The program is basically the same as the one in Project 17 except that the temperature high limit (TH) is set to 50 so that the THIGH output goes high when the temperature is equal to or exceeds 25°C and this turns on the buzzer to give a warning sound.

Program Listing

The full program listing is shown in Fig. 5.9. In addition to the program listing of Project 17, a function called *set_thigh* is added to load the temperature high limit. This function sends protocol number 01 to the DS1620 and then sends the data value 50 to set TH to 25°C.

Required Components

In addition to the components used for Project 17, a MOSFET transistor (e.g. VN66AFN) and a small buzzer will be required for this project.

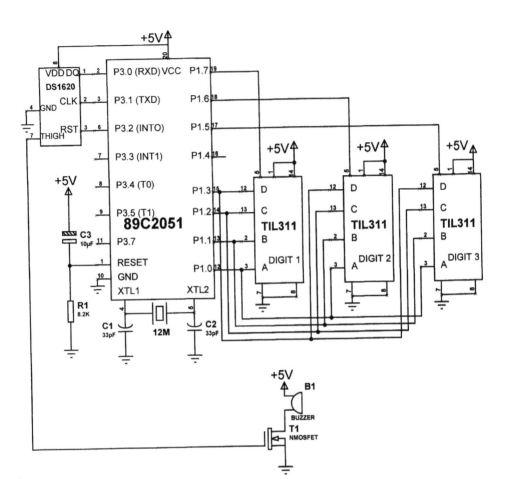

Figure 5.8.
Circuit diagram of Project 19

PROJECT 20 – Digital Thermometer With High and Low Alarm Outputs

Function

This project is similar to Project 19 except that the DS1620 is programmed so that a buzzer connected to the TCOM output of DS1620 turns on when the temperature is equal to or greater than TH and it then turns off only when the temperature drops below or equal to TL. In this project, TL is set to 25°C and TH is set to 30°C.

Temperature Projects

```
/********************************************************************
PROJECT:            PROJECT 19
FILE:               PROJ19.C
DATE:               August 1998
PROCESSOR:          AT892051
```

This is a temperature monitoring project. A DS1620 type digital thermometer is used to read the ambient temperature. The temperature is then displayed on three TL311 type alphanumeric displays. The temperature range is −55°C to +125°C. Positive temperature is displayed with leading zeros. Negative temperatures are displayed by inserting the letter 'E' in front of the display. The display accuracy is +/−1°C, i.e. there is no decimal point in the displayed data.

The display is updated every second.

A buzzer alarm is connected to the THIGH output of DS1620. DS1620 IC is loaded so that an alarm (buzzer) is generated when the temperature is above 25°C.
```
********************************************************************/

#include <AT892051.h>

sbit digit1_latch = P1^7;       /*digit 1 latch*/
sbit digit2_latch = P1^6;       /*digit 2 latch*/
sbit digit3_latch = P1^5;       /*digit 3 latch*/

sbit ds1620_dq = P3^0;          /*DS1620 data pin*/
sbit ds1620_clk = P3^1;         /*DS1620 clock pin*/
sbit ds1620_rst = P3^2;         /*DS1620 reset pin*/

#define read_temp 0xAA          /*read temp command*/
#define start_conv 0xEE         /*start conversion command*/
#define write_config 0x0C       /*write config command*/
#define write_thigh 0x01        /*write THIGH*/

/* Function to delay about a second */
void wait_a_second()
{
   unsigned int x;
   for(x=0;x<33000;x++);
}

/* Function to display data on three TIL311 displays. Negative temperature is displayed
with a leading 'E' character. Display range is −55 to +125. */
void display_temperature(unsigned int x)
{
   unsigned int s;

   int first,second,third;
```

```c
    if(x > 255)                 /*if negative*/
    {
      x=~x;                     /*complement temp*/
      x++;                      /*add 1 for 2s comp*/
      x=x & 0xFF;               /*extract lower 8 bits*/
      x=x/2;                    /*get real temp*/
      first=14;                 /*display leading `E'*/
      second=x/10;
      third=x-10*second;
    }
    else
    {
      x=x/2;                    /*temp is positive*/
      first=x/100;              /*extract digit data*/
      s=x-100*first;
      second=s/10;
      third=s-second*10;
    }

    P1=first | 0xE0;            /*Send digit1 data*/
    digit1_latch=0;             /*Latch the digit1 data*/
    digit1_latch=1;             /*Set digit1 latch on*/

    P1=second | 0xE0;           /*Send digit2 data*/
    digit2_latch=0;             /*Latch the digit2 data*/
    digit2_latch=1;             /*Set digit2 latch on*/

    P1=third | 0xE0;            /*Send digit3 data*/
    digit3_latch=0;             /*Latch digit3 data*/
    digit3_latch=1;             /*Set digit3 latch on*/
}

/* This function sends a data bit to DS1620 thermometer IC */
void write_ds1620_bit(unsigned char b)
{
    ds1620_dq=b;                /*send bit*/
    ds1620_clk=0;               /*set clock 0*/
    ds1620_clk=1;               /*set clock 1*/
    ds1620_dq=1;                /*set data 1*/
}

/* This function reads a data bit from DS1620 */
unsigned char read_ds1620_bit()
{
    unsigned char b;
```

Temperature Projects

```c
      ds1620_clk=0;           /*set clock 0*/
      b=ds1620_dq;            /*read a bit*/
      ds1620_clk=1;           /*set clock 1*/
      return (b);             /*return bit*/
}

/* This function writes data/configuration to DS1620 */
void write_to_ds1620(unsigned char ds1620_function,
                    unsigned char ds1620_data, unsigned char bit_count)
{
   unsigned char i,this_bit;

   ds1620_rst=1;              /*set reset to 1*/
   for(i=0;i<8;i++)           /*send function...*/
   {
      this_bit=ds1620_function >> i;
      this_bit=this_bit & 0x01;
      write_ds1620_bit(this_bit);
   }
   for(i=0;i<bit_count;i++)   /*send data...*/
   {
      this_bit=ds1620_data >> i;
      this_bit=this_bit & 0x01;
      write_ds1620_bit(this_bit);
   }
   ds1620_rst=0;
}                             /*set reset to 0*/

/* This function reads data/configuration from the DS1620 */
unsigned int read_from_ds1620(unsigned char ds1620_function,
                              unsigned char bit_count)
{
   unsigned char i,this_bit;
   unsigned int ds1620_data;

   ds1620_data=0;             /*initialize data*/
   ds1620_rst=1;              /*set reset to 1*/
   for(i=0;i<8;i++)
   {                          /*send function...*/
      this_bit=ds1620_function >> i;
      this_bit=this_bit & 0x01;
      write_ds1620_bit(this_bit);
   }                          /*read data*/
   for(i=0;i<bit_count;i++)
   {
```

```c
      ds1620_data=ds1620_data | read_ds1620_bit() << i;
   }
   ds1620_rst=0;
   return (ds1620_data);
}

/* This function configures the DS1620 for continuous operation */
void configure_ds1620()
{
   write_to_ds1620(write_config,2,8);
   wait_a_second();
}

/* This function starts the conversion */
void start_temp_conv()
{
   write_to_ds1620(start_conv,0,0);
}

/* This function reads the temperature */
unsigned int read_temperature()
{
   unsigned int t;

   t=read_from_ds1620(read_temp,9);    /*read temp*/
   return (t);                          /*return temp*/
}

/* This function writes to the THIGH register */
void set_thigh(int t)
{
   write_to_ds1620(write_thigh,t,9);
   wait_a_second();
}

/* Start of main program */
main()
{
   unsigned int TEMP;

   digit1_latch=1;            /*Set digit1 latch*/
   digit2_latch=1;            /*set digit2 latch*/
   digit3_latch=1;            /*set digit3 latch*/

   configure_ds1620();        /*configure DS1620*/
```

```
        set_thigh(50);              /*set THIGH for 25C*/
        start_temp_conv();          /*start conversion*/

        for(;;)                     /*endless loop*/
        {
           TEMP=read_temperature(); /*read temperature*/
           display_temperature(TEMP); /*Output to TIL311*/
           wait_a_second();         /*wait a second*/
        }
    }
```

Figure 5.9.

Program listing of Project 19

Circuit Diagram

The circuit diagram of this project is same as in Fig. 5.8 except that the MOSFET transistor is connected to pin 5 (TCOM) of the DS1620 instead of pin 7 (see block diagram, Fig. 5.10). A small buzzer is connected to the TCOM output of the DS1620 via a MOSFET power transistor. Normally the TCOM output is at logic low level and this output goes to logic high when the temperature exceeds the value TH (stored in the non-volatile memory of the DS1620) and then goes back to logic low when the temperature is equal to or less than TL.

Program Description

The program is basically the same the as the one in Project 19 except that the temperature high limit (TH) is set to 60 so that THIGH output goes high when

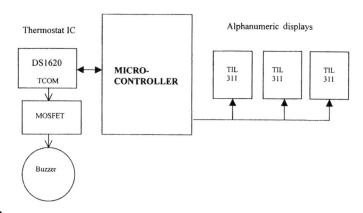

Figure 5.10.

Block diagram of Project 20

the temperature is equal to or exceeds 30°C and TL is set to 50 so that the TCOM output goes back to 0 when the temperature drops to 25°C or below, i.e. the buzzer will turn on when the temperature reaches 30°C and it will go off only when the temperature drops to 25°C or below.

Program Listing

The full program listing is shown in Fig. 5.11. This listing is the same as the listing in Project 19 with the addition of a function called *set_tlow* which sets the low limit (TL) of the DS1620.

Required Components

This project uses the same components as in Project 19.

PROJECT 21 – Using Analogue Temperature Sensor IC with A/D Converter

Function

This project shows how we can connect an analogue-to-digital (A/D) converter IC to our microcontroller. In this project, an analogue temperature sensor IC (LM35DZ) is used and its output is connected to an 8-bit A/D converter (ADC0804). The temperature is then displayed every second on a TSM5034 type 4-digit display. The block diagram of this project is shown in Fig. 5.12.

The A/D converter shown in this project can be connected to any kind of analogue voltage. For example, a digital voltmeter can be constructed easily by connecting the A/D converter input to an external voltage which is to be measured.

Circuit Diagram

Before looking at the circuit diagram of this project, it will be useful if we look at the ways an A/D converter can be connected to a microcontroller. There are many types of A/D converters available on the market. Some converters provide serial output data such that the output data is obtained from the converter each time a clock pulse is sent to the converter. These converters are very slow and are generally used where the speed of conversion is not important and where space saving is required. Serial A/D converters interface to a microcontroller by using only a few pins.

Standard A/D converters are generally used in medium- and high-speed applications. An example of such an A/D converter is the ADC0804,

Temperature Projects

/**
PROJECT: PROJECT 20
FILE: PROJ20.C
DATE: August 1999
PROCESSOR: AT892051

This is a temperature monitoring project. A DS1620 type digital thermometer is used to read the ambient temperature. The temperature is then displayed on three TL311 type alphanumeric displays. The temperature range is $-55°C$ to $+125°C$. Positive temperature is displayed with leading zeros. Negative temperatures are displayed by inserting the letter 'E' in front of the display. The display accuracy is $+/-1°C$. i.e. there is no decimal point in the displayed data.

The display is updated every second.

A buzzer alarm is connected to the TCOM output of DS1620. DS1620 IC is loaded so that an alarm (buzzer) is generated when the temperature is above $30°C$ and the alarm stops when the temperature drops below $25°C$, i.e. THIGH is loaded with $30°C$ and TLOW is loaded with $25°C$
**/

```
#include <AT892051.h>

sbit digit1_latch = P1^7;          /*digit 1 latch*/
sbit digit2_latch = P1^6;          /*digit 2 latch*/
sbit digit3_latch = P1^5;          /*digit 3 latch*/

sbit ds1620_dq = P3^0;             /*DS1620 data pin*/
sbit ds1620_clk = P3^1;            /*DS1620 clock pin*/
sbit ds1620_rst = P3^2;            /*DS1620 reset pin*/

#define read_temp 0xAA             /*read temp command*/
#define start_conv 0xEE            /*start conversion command*/
#define write_config 0x0C          /*write config command*/
#define write_thigh 0x01           /*write THIGH*/
#define write_tlow 0x02            /*write TLOW*/

/* Function to delay about a second */
void wait_a_second()
{
   unsigned int x;
   for(x=0;x<33000;x++);
}

/* Function to display data on three TIL311 displays. Negative temperature is displayed
with a leading 'E' character. Display range is -55 to +125. */
void display_temperature(unsigned int x)
{
   unsigned int s;
```

```c
      int first,second,third;

      if(x > 255)              /*if negative*/
      {
         x=~x;                 /*complement temp*/
         x++;                  /*add 1 for 2s comp*/
         x=x & 0xFF;           /*extract lower 8 bits*/
         x=x/2;                /*get real temp*/
         first=14;             /*display leading `E'*/
         second=x/10;
         third=x-10*second;
      }
      else
      {
         x=x/2;                /*temp is positive*/
         first=x/100;          /*extract digit data*/
         s=x-100*first;
         second=s/10;
         third=s-second*10;
      }
      P1=first | 0xE0;         /*Send digit1 data*/
      digit1_latch=0;          /*Latch the digit1 data*/
      digit1_latch=1;          /*Set digit1 latch on*/

      P1=second | 0xE0;        /*Send digit2 data*/
      digit2_latch=0;          /*Latch the digit2 data*/
      digit2_latch=1;          /*Set digit2 latch on*/

      P1=third | 0xE0;         /*Send digit3 data*/
      digit3_latch=0;          /*Latch digit3 data*/
      digit3_latch=1;          /*Set digit3 latch on*/
}

/* This function sends a data bit to DS1620 thermometer IC */
void write_ds1620_bit(unsigned char b)
{
      ds1620_dq=b;             /*send bit*/
      ds1620_clk=0;            /*set clock 0*/
      ds1620_clk=1;            /*set clock 1*/
      ds1620_dq=1;             /*set data 1*/
}

/* This function reads a data bit from DS1620 */
unsigned char read_ds1620_bit()
{
      unsigned char b;
```

Temperature Projects

```c
      ds1620_clk=0;              /*set clock 0*/
      b=ds1620_dq;               /*read a bit*/
      ds1620_clk=1;              /*set clock 1*/
      return (b);                /*return bit*/
}

/* This function writes data/configuration to DS1620 */
void write_to_ds1620(unsigned char ds1620_function,
                     unsigned char ds1620_data, unsigned char bit_count)
{
   unsigned char i,this_bit;

   ds1620_rst=1;                 /*set reset to 1*/
   for(i=0;i<8;i++)              /*send function...*/
   {
      this_bit=ds1620_function >> i;
      this_bit=this_bit & 0x01;
      write_ds1620_bit(this_bit);
   }
   for(i=0;i<bit_count;i++)      /*send data...*/
   {
      this_bit=ds1620_data >> i;
      this_bit=this_bit & 0x01;
      write_ds1620_bit(this_bit);
   }
   ds1620_rst=0;
}                                /*set reset to 0*/

/* This function reads data/configuration from the DS1620 */
unsigned int read_from_ds1620(unsigned char ds1620_function,
                              unsigned char bit_count)
{
   unsigned char i,this_bit;
   unsigned int ds1620_data;

   ds1620_data=0;                /*initialize data*/
   ds1620_rst=1;                 /*set reset to 1*/
   for(i=0;i<8;i++)
   {                             /*send function...*/
      this_bit=ds1620_function >> i;
      this_bit=this_bit & 0x01;
      write_ds1620_bit(this_bit);
   }                             /*read data*/
   for(i=0;i<bit_count;i++)
   {
```

```c
      ds1620_data=ds1620_data | read_ds1620_bit() << i;
   }
   ds1620_rst=0;
   return (ds1620_data);
}

/* This function configures the DS1620 for continuous operation */
void configure_ds1620()
{
   write_to_ds1620(write_config,2,8);
   wait_a_second();
}

/* This function starts the conversion */
void start_temp_conv()
{
   write_to_ds1620(start_conv,0,0);
}

/* This function reads the temperature */
unsigned int read_temperature()
{
   unsigned int t;

   t=read_from_ds1620(read_temp,9);    /*read temp*/
   return (t);                          /*return temp*/
}

/* This function writes to the THIGH register */
void set_thigh(int t)
{
   write_to_ds1620(write_thigh,t,9);
   wait_a_second();
}

/* This function writes to the TLOW register */
void set_tlow(int t)
{
   write_to_ds1620(write_tlow,t,9);
   wait_a_second();
}

/* Start of main program */
main()
{
```

```
unsigned int TEMP;

digit1_latch=1;              /*set digit1 latch*/
digit2_latch=1;              /*set digit2 latch*/
digit3_latch=1;              /*set digit3 latch*/

configure_ds1620();          /*configure DS1620*/
set_thigh(60);               /*set THIGH for 30C*/
set_tlow(50);                /*set TLOW for 25C*/
start_temp_conv();           /*start conversion*/

for(;;)                      /*endless loop*/
{
   TEMP=read_temperature();  /*read temperature*/
   display_temperature(TEMP);/*Output to TIL311*/
   wait_a_second();          /*wait a second*/
}
}
```

Figure 5.11.
Program listing of Project 20

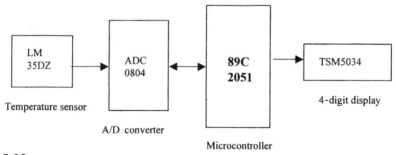

Figure 5.12.
Block diagram of Project 21

manufactured by the National Semiconductor Corporation. The conversion time of this A/D converter is 100 μs. As shown in Fig. 5.13, these converters interface to the microcontroller using the following pins (only the pins used in a standard application are shown):

DB0-DB7   8 data output pins
RD   Read input
WR   Write input
INTR   Interrupt output

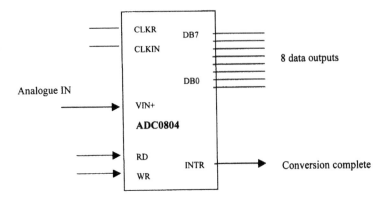

Figure 5.13.
ADC0804 Functional pin configuration

CLK R/CLK IN   Clock control inputs
VIN+   Positive analogue input

DB0 to DB7 are the digital output lines and the converted data appears on these eight lines. An 8-bit converter has 256 possible combinations (0 to 255) of output bit patterns. With a full-scale voltage of +5 V, the accuracy of the converter is $5/256 = 19.53$ mV. For example, a digital output pattern of '00010000' (i.e. decimal 16) corresponds to 312.48 mV. Similarly, a digital output pattern of '10100000' (i.e. decimal 160) corresponds to 3124.8 mV or 3.124 V and so on.

RD is the read data control pin and when RD is low (logic 0), output data appears on the eight output pins. When RD is high (logic 1), the output is not available.

WR input is normally at logic high and this input should be set to low and then high again for a conversion to start.

INTR is the interrupt output of the A/D converter. A high to low pulse is generated on this pin when a conversion is complete. This output is usually used to generate an interrupt in the microcontroller so that the converted data can be read.

ADC0804 contains an internal oscillator and it is required to connect an external resistor and a capacitor to pins CLK R and CLK IN to start the oscillator.

VIN+ is the pin where the analog input voltage should be applied.

Temperature Projects

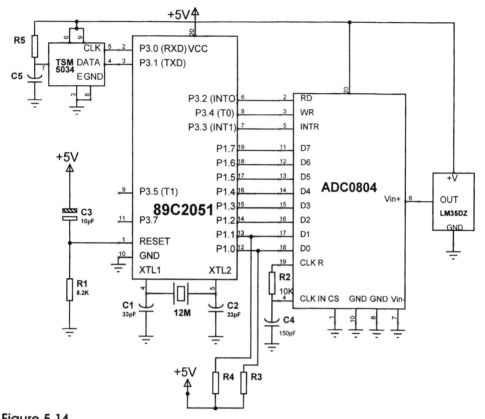

Figure 5.14.
Circuit diagram of Project 21

To make a single conversion the operation of the A/D converter can be summarized using the following steps:

- Set WR and RD high
- Start conversion by setting WR low
- Set WR back to high
- Detect end of conversion when INTR goes low (usually by interrupt)
- Set RD low and read data from DB0 to DB7
- Set RD high

The above process is of course repeated when more than one conversion is required.

Figure 5.14 shows the full circuit diagram of Project 21. Data and clock inputs of the TSM5034 are connected to bits 1 and 0 of port 3 respectively. The eight data outputs of the ADC0804 are connected to port 1. RD input is connected to bit 2 of port 3. WR is connected to bit 4 of port 3. The interrupt output, INTR, of the A/D converter is connected to bit 3 of port 3 which is the external interrupt 1 (INT1) pin of the microcontroller. Analogue input voltage is applied to pin 6 of the A/D converter and this input can be connected to any kind of analogue voltage which is to be measured.

In this project, analogue data comes from an LM35DZ type IC analogue temperature sensor. LM35DZ is a simple temperature sensor IC. Pin 1 of the device is connected to a power supply (e.g. +5 V), pin 3 is connected to the ground. Pin 2 is the output and this output provides a voltage which is directly proportional to the measured temperature. The device can measure temperatures from 2°C up to 100°C (some types can measure a wider range) and the output voltage to temperature relationship is 10 mV/°C. For example, at 20°C the output is 200 mV. Similarly, at 35°C, the output voltage is 350 mV, and so on. Internal oscillator of the A/D converter is activated by connecting an external resistor and a capacitor to pins CLK R and CLK IN. Notice that bits 0 and 1 of port 1 are connected to +5 V using pull-up resistors. This is necessary in some applications since the output drivers at these pins are open drain (i.e. there are no internal pull-up resistors).

Program Description

The display part of the program is as described in the *light projects* section of the book. We shall therefore look at the way the A/D converter is controlled by the software. The following PDL describes the operation of the project:

Main program

START
    Clear display
    Initialize microcontroller interrupts and A/D converter
    Start conversion
    DO FOREVER
    ENDDO
END

Initialization function

START
    Set A/D RD and WR pins to 1
    Set INT1 to accept interrupts on high-to-low edge
    Enable INT1 interrupts
    Set microcontroller to accept interrupts
END

Start conversion function

START
    Set WR pin to 0
    Set WR pin to 1
END

External interrupt INT1 service routine

START
    Set RD to 0
    Read temperature
    Convert to mV
    Set RD to 1
    Display temperature
    Wait a second
    Start conversion
END

The program clears the display and then initializes the microcontroller so that external interrupts on pin INT1 can be recognized. The A/D is then initialized and the conversion started. At the end of a conversion, an interrupt is generated by the A/D converter and execution jumps to the INT1 service routine. Here, the converter data is read and displayed on the TSM5034 display. At the same time a new conversion is restarted.

Program Listing

The complete program listing is shown in Fig. 5.15. Display clock and display data variables are assigned to pins 0 and 1, respectively, of port 3 of the microcontroller. Similarly, A/D RD and A/D WR variables are assigned to pins 2 and 4 of port 3 of the microcontroller. When the program starts it first clears the display. Function *initialize* is then called to set the A/D *RD* and *WR* inputs (*ADC_RD* and *ADC_WR*) to 1. External interrupt pin *INT1* of the microcontroller is also set ($IT1 = 1$) in this routine to accept interrupts on high-to-low transition and the microcontroller is configured to accept interrupts ($EA = 1$). Function *start_conversion* is then called to start an A/D conversion. This function simply sets the *WR* input of the A/D to 0 and then back to 1. The program then enters an endless loop and waits for external interrupts on its *INT1* pin.

INT1 has the interrupt number of 2. When a conversion is complete, control passes to the interrupt service routine *int1*. In this routine, *RD* input of the A/D converter is set to 0 to enable the output buffers and then the digital data is read into port 1 of the microcontroller. The value read is then converted to true

142 Microcontroller Projects in C for the 8051

```
/******************************************************************
PROJECT:              PROJECT 21
FILE:                 PROJ21.C
DATE:                 August 1999
PROCESSOR:            AT892051

This is a digital temperature, using an analogue-to-digital converter.

An ADC0804 type A/D converter is connected to port 1 of the microcontroller. Also, a
TSM5034 type 4-digit display is connected to port 3. The microcontroller controls both the
display and the A/D converter.

An LM35DZ type analogue temperature sensor IC is used to measure the temperature.
The voltage output of the LM35DZ is fed to the analogue input of the A/D converter.
Temperatures from 0°C up to 100°C, in steps of 0.5°C, are displayed on the 4-digit
display.
*******************************************************************/
#include <AT892051.h>

sbit DISPLAY_CLOCK=P3^0;       /*display clock*/
sbit DISPLAY_DATA =P3^1;       /*display data*/

sbit ADC_RD = P3^2;            /*A-D RD input*/
sbit ADC_WR = P3^4;            /*A-D WR input*/

unsigned int TEMPERATURE;      /*A-D data*/

/* This function provides a 1 second delay */
void wait_a_second()
{
  unsigned int x;
  for(x=0;x<33000;x++);
}

/* This function ends a clock pulse to the display */
void send_clock()
{
  DISPLAY_CLOCK=1;
  DISPLAY_CLOCK=0;
}

/* This function displays a digit */
void display_digit(int x,char dp)
{
  unsigned char LED_ARRAY[11]=
    {
      0xFC,0x60,0xDA,0xF2,0x66,0xB6,
      0xBE,0xE0,0xFE,0xF6,0
    };
```

Temperature Projects

```
   unsigned char n,top_bit,i;

   n=LED_ARRAY(x) | dp;
   for(i=1;i<=8;i++)
   {
      top_bit=n & 0x80;         /*Get top bit*/
      if(top_bit != 0)
         DISPLAY_DATA=1;
      else
         DISPLAY_DATA=0;
      send_clock();
      n=n << 1;                 /*Shift left by 1 digit*/
   }
}

/* This function displays all the 4 digits */
void display_all(int n)
{
   int r,first,second,third,fourth;

   first=n/1000;
   r=n-1000*first;
   second=r/100;
   r=r-100*second;
   third=r/10;
   fourth=r-third*10;

   DISPLAY_DATA=1;
   send_clock();

   if(n < 1000)                 /*Blank leading zeros*/
      display_digit(10,0);
   else
      display_digit(first,0);
   if(n < 100)
      display_digit(10,0);
   else
      display_digit(second,0);
   if(n < 10)
      display_digit(10,1);
   else
      display_digit(third,1);
   display_digit(fourth,0);

   DISPLAY_DATA=0;
```

```c
    send_clock();
    send_clock();
    send_clock();
}

/* This function clears the display */
void clear_display()
{
   int i;
   DISPLAY_DATA=0;
   DISPLAY_CLOCK=0;
   DISPLAY_DATA=1;
   send_clock();
   DISPLAY_DATA=0;
   for(i=1;i<=35;i++)send_clock();
}

/* This function initializes the A/D converter */
void initialize()
{
   ADC_RD=1;              /*set A-D RD to 1*/
   ADC_WR=1;              /*set A-D WR to 1*/
   IT1=1;                 /*set falling edge interrupt*/
   EX1=1;                 /*enable external INT1*/
   EA=1;                  /*enable interrupts*/
}

/* This function starts an A-D conversion */
void start_conversion()
{
   ADC_WR=0;              /*set A-D WR to 0*/
   ADC_WR=1;              /*set A-D WR to 1*/
}

/* This is the external interrupt INT1 service routine */
int1() interrupt 2
{
   ADC_RD=0;                            /*set RD to 0*/
   TEMPERATURE=P1;                      /*read A-D data*/
   TEMPERATURE=TEMPERATURE*19.60;       /*convert to true temp*/
   ADC_RD=1;                            /*set A-D RD to 1*/
   display_all(TEMPERATURE);            /*display the data*/
   wait_a_second();                     /*delay a bit*/
   start_conversion();                  /*display next conversion*/
}
```

```
/* Start of main program */
main()
{
  clear_display();        /*Clear display*/
  initialize();           /*initialize A-D*/
  start_conversion();     /*start conversion*/

  for(;;)                 /*endless loop*/
  {
  }
}
```

Figure 5.15.

Program listing of Project 21

temperature and displayed by calling the function *display_all*. After about a second delay, a new conversion is started and the above process repeats.

Required Components

In addition to the standard components required by the basic microcontroller circuit, the following components are required for this project:

ADC0804	A/D converter IC
TSM5034	Display IC
LM35DZ	Temperature sensor IC
R2	10K
R3, R4	100K
R5	8.2K resistor
C4	150 pF capacitor
C5	0.1 µF capacitor

All resistors are 0.125 W.

CHAPTER 6

RS232 SERIAL COMMUNICATION PROJECTS

RS232 is a serial communications standard which enables data to be transferred in serial form between two devices. Data is transmitted and received in serial 'bit stream' from one point to another. Standard RS232 is suitable for data transfer to about 50 m, although special low-loss cables can be used for extended distance operation. Four parameters specify an RS232 link between two devices. These are *baud rate*, *data width*, *parity*, and the *stop bits*, and are described below:

- *Baud rate*: the baud rate (bits per second) determines how much information is transferred over a given time interval. A baud rate can usually be selected between 110 and 76 800 baud, e.g. a baud rate of 9600 corresponds to 9600 bits per second.
- *Data width*: the data width can be either 8 bits or 7 bits depending upon the nature of the data being transferred.
- *Parity*: the parity bit is used to check the correctness of the transmitted or received data. Parity can either be even, odd, or no parity bit can be specified at all.
- *Stop bit*: the stop bit is used as the terminator bit and it is possible to specify either one or two stop bits.

Serial data is transmitted and received in frames where a frame consists of:

- 1 start bit
- 7 or 8 data bits
- optional parity bit
- 1 stop bit

In many applications 10 bits are used to specify an RS232 frame, consisting of 1 start bit, 8 data bits, no parity bit, and 1 stop bit. For example, character 'A' has the ASCII bit pattern '01000001' and is transmitted as shown in Fig. 6.1 with 1 start bit, 8 data bits, no parity, and 1 stop bit. The data is transmitted least significant bit first.

When 10 bits are used to specify the frame length, the time taken to transmit or receive each bit can be found from the baud rate used. Table 6.1 gives the time

148 Microcontroller Projects in C for the 8051

Figure 6.1.
Transmitting character 'A' (bit pattern 01000001)

Table 6.1	Bit times for different baud rates
Baud rate	**Bit time**
300	3.33 ms
600	1.66 ms
1200	833 µs
2400	416 µs
4800	208 µs
9600	104 µs
19 200	52 µs

taken for each bit to be transmitted or received for most commonly used baud rates.

RS232 Connectors

As shown in Fig. 6.2, two types of connectors are used for RS232 communications. These are the 25-way D-type connector (known as DB25) and the 9-pin D-type connector (also known as DB9). Table 6.2 lists the most commonly used signal names for both DB9 and DB25 type connectors. The used signals are:

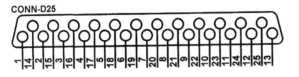
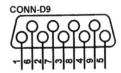

Figure 6.2.
RS232 connectors

Table 6.2 Commonly used RS232 signals

Description	Signal	9 pin	25 pin
Carrier detect	CD	1	8
Receive data	RD	2	3
Transmit data	TD	3	2
Data terminal ready	DTR	4	20
Signal ground	SG	5	7
Data set ready	DSR	6	6
Request to send	RTS	7	4
Clear to send	CTS	8	5
Ring indicator	RI	9	22

SG: signal ground. This pin is used in all RS232 cables.
RD: received data. Data is received at this pin. This pin is used in all two-way communications.
TD: transmit data. Data is sent out from this pin. This pin is used in all two-way communications.
RTS: request to send. This signal is asserted when the device requests data to be sent.
CTS: clear to send. This signal is asserted when the device is ready to accept data.
DTR: data terminal ready. This signal is asserted to indicate that the device is ready.
DSR: data set ready. This signal indicates, by the device at the other end, that it is ready.
CD: carrier detect. This signal indicates that a carrier signal has been detected by a modem connected to the line.

In some RS232 applications it is sufficient to use only the pins SG, RD, and TD. Also, in some applications (e.g. when two similar devices are connected together) it is necessary to twist pins RD and TD so that the transmit pin of one device is connected to the receive pin of the other device and vice versa.

RS232 Signal Levels

RS232 is bi-polar and a voltage of $+3$ to $+12$ V indicates an ON state (or SPACE), while a voltage of -3 to -12 V indicates an OFF state (or MARK). In practice, the ON and OFF states can be achieved with lower voltages.

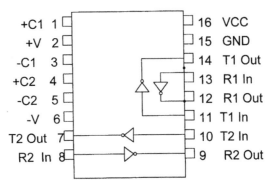

Figure 6.3.
Pin configuration of MAX232

Standard TTL logic devices, including the 89C2051 microcontroller, operate with TTL logic levels between the voltages of 0 and +5 V. Voltage level converter ICs are used to convert between the TTL and RS232 voltage levels. One such popular IC is the MAX232, manufactured by MAXIM, and operaters with +5 V supply. The MAX232 is a 16-pin DIL chip incorporating two receivers and two transmitters (see Fig. 6.3) and the device requires four external capacitors for proper operation.

The 89C2051 microcontroller can output TTL level RS232 signals from its TXD (or pin P3.1) pin and it can receive TTL level RS232 signals from its RXD (or pin P3.0) pin. The microcontroller can be connected to external RS232 compatible equipment via a MAX232 type voltage converter IC.

Controlling the RS232 Port

Before the serial port can be used it is necessary to set various registers:

SCON: this is the serial port control register. It should be set to hexadecimal 0x50 for 8-bit data mode.

TMOD: this register controls the timers for baud rate generation and it should be set to hexadecimal 0x20 to enable timer 1 to operate in 8-bit auto-reload mode.

TH1: this register should be loaded with a constant so that the required baud rate can be generated. Table 6.3 shows the values to be loaded into TH1 and the corresponding baud rates for two different clock rates.

TR1: this register starts/stops the timer and it should be set to 1 to start timer 1.

TI: this register should be set to 1 to indicate ready to transmit.

RS232 Serial Communication Projects 151

Table 6.3 TH1 values for different baud rates

Baud rate	Clock	SMOD	TH1 value	Error
9600	12.000 MHz	1	0xF9	7%
4800	12.000 MHz	0	0xF9	7%
2400	12.000 MHz	0	0xF3	0.16%
1200	12.000 MHz	0	0xE6	0.16%
9600	11.059 MHz	0	0xFD	0
4800	11.059 MHz	0	0xFA	0
2400	11.059 MHz	0	0xF4	0
1200	11.059 MHz	0	0xE8	0

Note that register SMOD should be set to 1 when we require 9600 baud at 12 MHz clock rate. SMOD is set to 0 at reset time.

For example, the following function shows how we can initialize the serial port for 2400 baud operation:

```
void serial_init()
{
    SCON=0x50;
    TMOD=0x20;
    TH1=0xF3;
    TR1=1;
    TI=1;
}
```

PROJECT 22 – Output a Simple Text Message from the RS232 Port

Function

This project shows how we can interface our microcontroller to an external RS232 compatible device (e.g. an RS232 visual display unit, or COM1 or COM2 port of a PC) and send a text message to this device. The text message 'THIS IS AN RS232 TEST MESSAGE' is sent out continuously from the microcontroller. The frame format used in this project is 2400 baud, 8 data bits, no parity, and 1 stop bit.

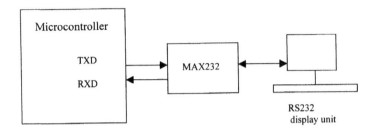

Figure 6.4.
Block diagram of Project 22

Circuit Diagram

The block diagram of Project 22 is shown in Fig. 6.4. The TXD pin of the microcontroller is connected to the MAX232 Maxim voltage converter IC and the output of this IC can be connected to the input of a COM1 (or COM2) port of a PC, or to the input of an RS232 visual display unit. Similarly, the output of the external RS232 device is connected to the RXD input of the microcontroller via the MAX232 IC. A terminal emulation software can be activated on the PC to receive and display any data arriving at its serial port.

The complete circuit diagram of this project is shown in Fig. 6.5. Pin P3.1 of the microcontroller (TXD) is connected to pin 10 of the MAX232 converter IC. Pin 7 of this IC is connected to the external RS232 compatible serial device which is to receive and display our text message. Similarly, the output of the RS232 device is connected to pin 8 input of the MAX232 IC and pin 9 output of this IC is connected to pin 2 (RXD) serial input of the microcontroller. Correct operation of MAX232 requires four external capacitors to be connected as shown in the figure.

Program Description

The program initializes the RS232 port of the microcontroller and then sends a test message to the port.

The following PDL describes the functions of the program:

START
    Initialize RS232 port
    DO FOREVER
        Display text 'THIS IS AN RS232 TEST MESSAGE'
    ENDDO
END

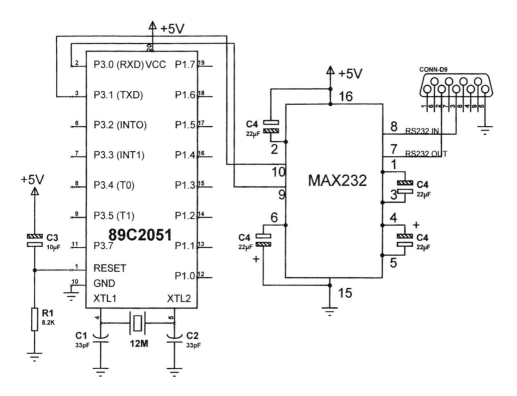

Figure 6.5.
Circuit diagram of Project 22

Program Listing

The program listing is given in Fig. 6.6. Notice that the standard input/output library 'stdio.h' is included at the top of the program. The main program calls function *serial_init* which initializes the RS232 port to 2400 baud and enables transmissions. Standard C function *printf* is used to send the serial data to the RS232 port. A carriage return and line feed pair ('\n') are sent after each output.

It is important to notice that this simple program occupies about 1094 bytes in the memory of the microcontroller. This is because the *printf* function is a complex library function and is implemented in a large number of instructions. A simple function can be developed to emulate some of the functionalities of *printf* so that the output operations can be performed with less memory as described below.

```
/*************************************************************
PROJECT:              PROJECT 22
FILE:                 PROJ22.C
DATE:                 August 1999
PROCESSOR:            AT892051

This project sends the text message: 'THIS IS AN RS232 TEST MESSAGE' to the RS232 serial
port of the microcontroller. The message is sent out continuously with a carriage return
and line feed at the end of each line.

The program occupies about 1094 bytes of memory
*************************************************************/
#include <stdio.h>
#include <AT892051.h>

/* Function to initialize the RS232 serial port */
void serial_init()
{
   SCON=0x50;              /* setup for 8-bit data */
   TMOD=0x20;              /* setup timer 1 for auto-reload */
   TH1=0xF3;               /* setup for 2400 baud */
   TR1=1;                  /* turn on timer 1 */
   TI=1;                   /* indicate ready to transmit */
}

/* Start of main program */
main()
{
   serial_init();          /*initialize serial port*/
   for(;;)                 /*Start of loop*/
   {
      printf('THIS IS AN RS232 TEST MESSAGE\n');
   }
}
```

Figure 6.6.

Program listing of Project 22

A Simple Serial Output Function

The 89C2051 microcontroller is equipped with 2 Kbytes of memory. It was shown in the previous example that using the built-in *printf* function causes a large part of this memory to be used, leaving little space for other operations. Figure 6.7 shows a program listing that performs serial output functions without using the *printf* function and the complete program occupies about 400 bytes of memory. In this program, the serial transmit register of the microcontroller (SBUF) is used to send out data directly. Function *send_serial* transmits a null-terminated string to the RS232 port of the microcontroller. The program waits until the transmit register is empty (TI = 1) before sending out the next character. In this example, the string 'ANOTHER TEST' is output continuously. Notice that calling this function with variable *crlf* causes a carriage return and line feed to be output at the end of the test message.

Components Required

The following components will be required in addition to the standard microcontroller components:

MAX232 IC
C4   22 μF capacitor (4 off)
9 way or 25 way RS232 connector

PROJECT 23 – Input/Output Example Using the RS232 Port

Function

This project shows how we can input and output serial data using the built-in C functions. In this example, the user is prompted to enter a character through the RS232 terminal. The program then finds the next character (i.e. increments the character by one) and outputs it to the user's terminal.

Circuit Diagram

The circuit diagram of this project is the same as in Project 22 (i.e. Fig. 6.5).

Program Description

The RS232 serial port is initialized to operate at 2400 baud. The user is then prompted to enter a character. This character is incremented by one and sent to the serial output port.

The following PDL describes the functions of the program:

START
    Initialize serial port
    DO FOREVER
        Display 'Enter a character'
        Read a character
        Increment the character
        Display the next character
    ENDDO
END

```
/****************************************************************************
PROJECT:              PROJECT 22
FILE:                 PROJ22-1.C
DATE:                 August 1999
PROCESSOR:            AT892051

This project sends the text message: 'ANOTHER TEST' to the RS232 serial port of the
microcontroller. The message is sent out continuously with a carriage return and line
feed at the end of each line.

This program does not use the built-in function printf. The program occupies about 400
bytes of memory.
*****************************************************************************/
#include <stdio.h>
#include <AT892051.h>

/* Function to initialize the RS232 serial port */
void serial_init()
{
   SCON=0x50;              /* setup for 8-bit data */
   TMOD=0x20;              /* setup timer 1 for auto-reload */
   TH1=0xF3;               /* setup for 2400 baud */
   TR1=1;                  /* turn on timer 1 */
   TI=1;                   /* indicate ready to transmit */
}

/* This function displays a null-terminated string on the RS232 port */
void send_serial(unsigned char *s)
{
   while(*s != 0x0)
   {
      SBUF=*s;             /*send out the character*/
      While(! TI)          /*wait until sent*/
      {
      }
      TI=0;
      s++;                 /*get the next character*/
   }
}

/* Start of main program */
main()
{
   unsigned char crlf[ ]={0x0D,0x0A,0x0};  /*carriage return, line feed*/
   serial_init();                          /*initialize serial port*/
   for(;;)                                 /*Start of loop*/
```

```
    {
        send_serial('ANOTHER TEST');
        send_serial(crlf);
    }
}
```

Figure 6.7.
Output program which does not use the *printf* function

Program Listing

The program listing is shown in Fig. 6.8. Function *serial_init* initializes the serial port for operation at 2400 baud with a 12 MHz crystal. Built-in function *printf* is used to prompt the user to enter a character. A character is then read from the user's terminal using the standard C built-in function *getchar* and this character is stored in a variable called *c*. Finally, this character is incremented by one and is output to the RS232 port using function *printf*. The above process is repeated indefinitely. This program occupies 1164 bytes of memory.

Input/Output Without Using the Built-in Functions

The above program uses the standard C built-in functions *printf* and *getchar*. As a result the program is big. An example program is given in Fig. 6.9 which does not use these built-in functions and thus occupies much less space in memory.

Function *serial_init* is the same as before but note that the serial port interrupts are enabled (EA = 1 and ES = 1). Function *send_serial* sends a null-terminated string to the serial output port. Similarly, function *send_1_char* sends a single character to the serial port. Serial data is read in via the serial port interrupt service routine (*serial*). Whenever a character is transmitted or received, the interrupt service routine is activated automatically. The interrupt number of the serial port is 4. Here, the receive interrupt register (RI) is checked and a character is assumed to be received from the serial port if RI is non-zero. The received character is copied from SBUF to a variable called *received_character*.

The main program calls function *send_serial* to display the message 'Enter a character'. If a character is received, this character is echoed on the user's terminal and the next character is displayed by incrementing and outputting the variable *received_character*. Function *send_1_char* is then used to send a carriage return and line feed after each output.

```
/****************************************************************
PROJECT:                PROJECT 23
FILE:                   PROJ23.C
DATE:                   August 1999
PROCESSOR:              AT892051

This project is an example of using both the input and the output serial data routines. A
character is received from the serial port. The next character is then calculated and
output to the user's RS232 terminal.

This program occupies 1164 bytes in memory.
****************************************************************/
#include <stdio.h>
#include <AT892051.h>

/* Function to initialize the RS232 serial port */
void serial_init()
{
   SCON=0x50;               /* setup for 8-bit data */
   TMOD=0x20;               /* setup timer 1 for auto-reload */
   TH1=0xF3;                /* setup for 2400 baud */
   TR1=1;                   /* turn on timer 1 */
   TI=1;                    /* indicate ready to transmit */
}

/* Start of main program */
main()
{
   char c;
   serial_init();           /*initialize serial port*/

   for(;;)                  /*Start of loop*/
   {
     printf('\nEnter a character');
     c=getchar();           /*read a character*/
     c++;                   /*next character*/
     printf('The next character is: %c:,c);
   }
}
```

Figure 6.8.

Program listing of Project 23

RS232 Serial Communication Projects 159

```
/*******************************************************************************
PROJECT:                PROJECT 23
FILE:                   PROJ23-1.C
DATE:                   August 1999
PROCESSOR:              AT892051

This project reads a character from the user's terminal, finds the next character and
displays on the user's terminal. C built-in functions are not used in this program.

This program occupies 225 bytes of memory.
*******************************************************************************/
#include <stdio.h>
#include <AT892051.h>

unsigned char received_character;
int received_flag;

/* Function to initialize the RS232 serial port */
void serial_init()
{
   SCON=0x50;               /* setup for 8-bit data */
   TMOD=0x20;               /* setup timer 1 for auto-reload */
   TH1=0xF3;                /* setup for 2400 baud */
   TR1=1;                   /* turn on timer 1 */
   TI=1;                    /* indicate ready to transmit */
   EA=1;                    /*enable interrupts*/
   ES=1;                    /*enable serial port interrupts*/
}

/* This function displays a null-terminated string on the RS232 port */
void send_serial(unsigned char *s)
{
   while(*s != 0x0)
   {
      SBUF=*s;              /*send out the character*/
      while(! TI)           /*wait until sent*/
      {
      }
      TI=0;
      s++;                  /*get the next character*/
   }
}

/* This function sends a single character to the serial port */
void send_1_char(unsigned char c)
```

```
{
   SBUF=c;                          /*send out the character*/
   while(! TI)                      /*wait until transmitted*/
   {
   }
   TI=0;
}

/* Serial port interrupt service routine. Program jumps to this routine when a character is
transmitted or received */
serial() interrupt 4
{
   if(RI)                           /*if a character received*/
   {
      received_character=SBUF;
      RI=0;
      received_flag=1;              /set received flag*/
   }
}

/* Start of main program */
main()
{
   received_flag=0;
   serial_init();                   /*initialize serial port*/

   for(;;)                          /*Start of loop*/
   {
      send_serial('Enter a character:');
      while(received_flag == 0)
      {
      }
      received_flag=0;
      send_1_char(received_character);     /*echo*/
      send_serial('The next character is:');
      received_character++;                /*next char*/
      send_1_char(received_character);
      send_1_char(0x0D);            /*send carriage return*/
      send_1_char(0x0A);            /*send line feed*/
   }
}
```

Figure 6.9.

Program not using the built-in C functions

PROJECT 24 – A Simple Calculator Program Using the RS232 Port

Function

This is a simple calculator project based upon the 89C51 type microcontroller. The microcontroller is connected to an RS232 serial terminal. The user can perform simple addition, subtraction, multiplication, and division of numbers using the microcontroller.

Circuit Diagram

This project is based upon the 89C51 microcontroller. This is a 40-pin device which is software compatible with the 89C2051 microcontroller. The 89C51 contains a 4 Kbyte flash program memory, 128 bytes of RAM, 32 programmable input/output lines, and six interrupt sources.

The circuit diagram of this project is shown in Fig. 6.10. A 12 MHz crystal and two capacitors are connected to pins 18 and 19 of the microcontroller. Reset input is connected to a capacitor and a resistor. Transmit output (TXD) and receive input (RXD) of the device are connected to a MAX232 type RS232 converter IC. EA is the external program enable pin and this pin should be connected to +5 V for internal program executions.

Program Description

The program operates as a simple calculator. When power is applied to the microcontroller, a menu is displayed on the user's terminal and the user is prompted to enter two numbers and the operation to be performed. A typical dialogue is given below (note that the characters typed by the user are underlined for clarity):

A SIMPLE MICROCONTROLLER-BASED CALCULATOR
===
Enter 2 integer numbers and the operation
to be performed. Valid operations are:
+   ADD
−   SUBTRACT
*   MULTIPLY
/   DIVIDE
Enter First Number: 5
Enter Second Number: 3
Enter Operation: +
Result = 8
A SIMPLE ...

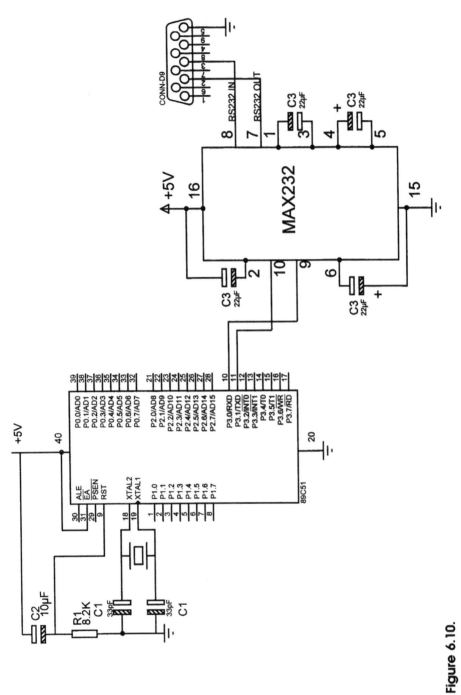

Figure 6.10.
Circuit diagram of Project 24

The following PDL describes the operation of the program:

START
  Initialize serial port
  Display heading
  Get a number 1
  Get number 2
  WHILE Operation is not valid
    Get Operation to be performed
  WEND
  IF Operation = '+'
    ADD the numbers
  ELSE IF Operation = '−'
    SUBTRACT the numbers
  ELSE IF Operation = '*'
    MULTIPLY the numbers
  ELSE IF Operation = '/'
    DIVIDE the numbers
  ENDIF
  Display the result
END

Program Listing

The program listing is given in Fig. 6.11. The serial port is initialized by the function *serial_init*. The program then prints a heading and a menu using the built-in function *printf*. The user is prompted to enter the numbers and the operation to be performed. The first number is stored in variable *num1* using the built-in function *scanf*. The second number is stored in variable *num2*. The operation to be performed is stored in variable *oper*. A *switch* statement is then used to select the required operation. The result of the calculation is stored in variable *result* and this is then displayed using a *printf* function. The program repeats forever unless stopped by the user.

Components Required

The following components will be required for this project:

89C51 microcontroller
MAX232 IC
12 MHz crystal
 C1 33 µF capacitors (2 off)
 C2 10 µF capacitor
 C3 22 µF capacitors (4 off)
 R1 8.2K resistor (0.125 W)
9 way or 25 way RS232 connector

```
/******************************************************************
PROJECT:              PROJECT 24
FILE:                 PROJ24.C
DATE:                 August 1999
PROCESSOR:            AT892051

This is a simple calculator project based upon the 89C51 type 40-pin microcontroller. The
project enables the user to perform integer addition, subtraction, multiplication, and
division.

This program occupies just over 2720 bytes of memory.
******************************************************************/
#include <stdio.h>
#include <AT892051.h>

/* Function to initialize the RS232 serial port */
void serial_init()
{
    SCON=0x50;              /* setup for 8-bit data */
    TMOD=0x20;              /* setup timer 1 for auto-reload */
    TH1=0xF3;               /* setup for 2400 baud */
    TR1=1;                  /* turn on timer 1 */
    TI=1;                   /* indicate ready to transmit */
}

/* Start of main program */
main()
{
    int num1,num2,result;
    char c,oper;
    serial_init();          /*initialize serial port*/

    for(;;)                 /*Start of loop*/
    {
        printf('\n\nA SIMPLE MICROCONTROLLER BASED CALCULATOR\n');
        printf('=============================================');
        printf('Enter 2 integer numbers and the operation\n');
        printf('to be performed. Valid operations are:\n');
        printf('+ ADD\n');
        printf('- SUBTRACT\n');
        printf('* MULTIPLY\n');
        printf(/ DIVIDE\n\n');
        printf('Enter First Number:');
        scanf('%d',&num1);
        c=gethar();
        printf('Enter Second Number:');
```

```
          scanf('%d,&num2');
          c=getchar();
          oper=' ';
          while(oper != '+' && oper != '-' && oper != '*' && oper != '/')
          {
            printf('Enter Operation:');
            oper=getchar();
            printf('\n');
          }
          switch (oper)
          {
            case '+':
              result=num1+num2;
              break;
            case '-':
              result=num1-num2;
              break;
            case '*':
              resut=num1*num2;
              break;
            case '/':
              result=num1/num2;
              break;
          }
          printf('Result = %d\n',result);
       }
    }
```

Figure 6.11.

Program listing of Project 24

Appendix

ASCII CODE

ASCII codes of the first 128 characters are standard and the same code is used between different equipment manufacturers. ASCII codes of characters between 128 and 255 are also known as the extended ASCII characters and these characters and their codes may differ between different computer manufacturers. Below is a list of the most commonly used ASCII characters and their codes both in hexadecimal and in binary.

Character	Hex	Binary	Character	Hex	Binary
NUL	00	00000000	EM	19	00011001
SOH	01	00000001	SUB	1A	00011010
STX	02	00000010	ESC	1B	00011011
ETX	03	00000110	FS	1C	00011100
EOT	04	00000100	GS	1D	00011101
ENQ	05	00000101	RS	1E	00011110
ACK	06	00000110	US	1F	00011111
BEL	07	00000111	SP	20	00100000
BS	08	00001000	!	21	00100001
HT	09	00001001	"	22	00100010
LF	0A	00001010	#	23	00100011
VT	0B	00001011	$	24	00100100
FF	0C	00001100	%	25	00100101
CR	0D	00001101	&	26	00100110
SO	0E	00001110	'	27	00100111
SI	0F	00001111	(	28	00101000
DLE	10	00010000	)	29	00101001
XON	11	00010001	*	2A	00101010
DC2	12	00010010	+	2B	00101011
XOFF	13	00010110	,	2C	00101100
DC4	14	00010100	-	2D	00101101
NAK	15	00010101	.	2E	00101110
SYN	16	00010110	/	2F	00101111
ETB	17	00010111	0	30	00110000
CAN	18	00011000	1	31	00110001

Character	Hex	Binary	Character	Hex	Binary
2	32	00110010	a	61	01100001
3	33	00110011	b	62	01100010
4	34	00110100	c	63	01100011
5	35	00110101	d	64	01100100
6	36	00110110	e	65	01100101
7	37	00110111	f	66	01100110
8	38	00111000	g	67	01100111
9	39	00111001	h	68	01101000
:	3A	00111010	i	69	01101001
;	3B	00111011	j	6A	01101010
<	3C	00111100	k	6B	01101011
=	3D	00111101	l	6C	01101100
>	3E	00111110	m	6D	01101101
?	3F	00111111	n	6E	01101110
@	40	01000000	o	6F	01101111
A	41	01000001	p	70	01110000
B	42	01000010	q	71	01110001
C	43	01000011	r	72	01110010
D	44	01000100	s	73	01110011
E	45	01000101	t	74	01110100
F	46	01000110	u	75	01110101
G	47	01000111	v	76	01110110
H	48	01001000	w	77	01110111
I	49	01001001	x	78	01111000
J	4A	01001010	y	79	01111001
K	4B	01001011	z	7A	01111010
L	4C	01001100	{	7B	01111011
M	4D	01001101	|	7C	01111100
N	4E	01001110	}	7D	01111101
O	4F	01001111	~	7E	01111110
P	50	01010000		7F	01111111
Q	51	01010001		80	10000000
R	52	01010010		81	10000001
S	53	01010011	'	82	10000010
T	54	01010100	ƒ	83	10000011
U	55	01010101	"	84	10000100
V	56	01010110	...	85	10000101
W	57	01010111	†	86	10000110
X	58	01011000	‡	87	10000111
Y	59	01011001	^	88	10001000
Z	5A	01011010	‰	89	10001001
[	5B	01011011	Š	8A	10001010
\	5C	01011100	<	8B	10001011
]	5D	01011101	Œ	8C	10001100
^	5E	01011110		8D	10001101
	5F	01011111		8E	10001110
`	60	01100000		8F	10001111

Glossary 169

Character	Hex	Binary	Character	Hex	Binary
	90	10010000	¾	BE	10111110
`	91	10010001	¿	BF	10111111
´	92	10010010	À	C0	11000000
ˏ	93	10010011	Á	C1	11000001
"	94	10010100	Â	C2	11000010
•	95	10010101	Ã	C3	11000011
–	96	10010110	Ä	C4	11000100
—	97	10010111	Å	C5	11000101
~	98	10011060	Æ	C6	11000110
™	99	10011001	Ç	C7	11000111
Š	9A	10011010	È	C8	11001000
>	9B	10011011	É	C9	11001001
œ	9C	10011100	Ê	CA	11001010
	9D	10011101	Ë	CB	11001011
	9E	10011110	Ì	CC	11001100
Ÿ	9F	10011111	Í	CD	11001101
	A0	10100000	Î	CE	11001110
¡	A1	10100001	Ï	CF	11001111
¢	A2	10100010	Ð	D0	11010000
£	A3	10100011	Ñ	D1	11010001
¤	A4	10100100	Ò	D2	11010010
¥	A5	10100101	Ó	D3	11010011
¦	A6	10100110	Ô	D4	11010100
§	A7	10100111	Õ	D5	11010101
¨	A8	10101000	Ö	D6	11010110
©	A9	10101001	×	D7	11010111
ª	AA	10101010	Ø	D8	11011000
«	AB	10101011	Ù	D9	11011001
¬	AC	10101100	Ú	DA	11011010
-	AD	10101101	Û	DB	11011011
®	AE	10101110	Ü	DC	11011100
¯	AF	10101111	Ý	DD	11011101
°	B0	10110000	Þ	DE	11011110
±	B1	10110001	ß	DF	11011111
²	B2	10110010	à	E0	11100000
³	B3	10110011	á	E1	11100001
´	B4	10110100	â	E2	11100010
µ	B5	10110101	ã	E3	11100011
¶	B6	10110110	ä	E4	11100100
·	B7	10110111	å	E5	11100101
¸	B8	10111000	æ	E6	11100110
¹	B9	10111001	ç	E7	11100111
º	BA	10111010	è	E8	11101000
»	BB	10111011	é	E9	11101001
¼	BC	10111100	ê	EA	11101010
½	BD	10111101	ë	EB	11101011

Character	Hex	Binary	Character	Hex	Binary
ì	EC	11101100	ö	F6	11110110
í	ED	11101101	÷	F7	11110111
î	EE	11101110	ø	F8	11111000
ï	EF	11101111	ù	F9	11111001
ð	F0	11110000	ú	FA	11111010
ñ	F1	11110001	û	FB	11111011
ò	F2	11110010	ü	FC	11111100
ó	F3	11110011	ý	FD	11111101
ô	F4	11110100	þ	FE	11111110
õ	F5	11110101	ÿ	FF	11111111

Glossary

ADC Analogue-to-digital converter. A device that converts analogue signals to a digital form for use by a computer.

Algorithm A fixed step-by-step procedure for finding a solution to a problem.

ANSI American National Standards Institute.

Architecture The arrangement of functional blocks in a computer system.

ASCII American Standard Code for Information Interchange. A widely used code in which alphanumeric characters and certain other special characters are represented by unique 7-bit binary numbers. For example, the ASCII code of the letter 'A' is 65.

Assembler A software that translates symbolically represented instructions into their binary equivalents.

Assembly language A source language that is made up of the symbolic machine language statements. Assembly language is very efficient since there is a one-to-one correspondence with the instruction formats and data formats of the computer.

BASIC Beginners All-purpose Symbolic Instruction Code. A high-level programming language commonly used in personal computers. BASIC is usually an interpreted language.

Baud The unit of data transmission speed. Baud is often equated to the number of serial bits transferred per second.

Baud rate Measurement of data flow in a serial communication system. Baud rate is typically equal to bits per second. Some typical baud rates are 9600, 4800, 2400 and so on.

BCD Binary Coded Decimal. A code in which each decimal digit is binary coded into 4-bit words. By representing binary numbers in this form, it is readily possible to display and print numbers.

Bi-directional port An interface port that can be used to transfer data in either direction.

Binary The representation of numbers in a base two system.

Bit A single binary digit.

Byte A group of 8 binary digits.

Chip A small rectangle of silicon on which an integrated circuit is fabricated.

Clock A circuit generating regular timing signals for a digital logic system. In microcomputer systems clocks are usually generated by using crystal devices. A typical clock frequency is 12 MHz.

CMOS Complementary Metal Oxide Semiconductor. A family of integrated circuits that offers extremely high packing density and low power.

Compiler A program designed to translate high-level languages into machine code.

Counter A register or a memory location used to record numbers of events as they occur.

CRT Cathode Ray Tube. A display screen.

Cycle time The time required to access a memory location or to carry out an operation in a computer system.

DAC Digital-to-analogue converter. A device that converts digital signals into analogue form.

Decimal system Base 10 numbering system.

Development system Equipment used to develop microprocessor- and microcomputer-based software and hardware projects.

Dot matrix Method of printing or displaying characters in which each character is formed by a rectangular array of dots to give the required shape.

EAROM Electrically Alterable Read Only Memory. In this type of memory part or all of the data can be erased and rewritten by applying electrical signals.

Edge triggered Circuit action initiated by the change of a signal. An edge could be the change of a signal from 0 to 1 or from 1 to 0.

Emulator Software or hardware system that duplicates the actions of a microprocessor or a microcomputer system.

EPROM Erasable Programmable Read Only Memory. This type of memory can be erased by exposure to ultraviolet light and then reprogrammed using a programmer.

Execute To perform a specified operational sequence in a program.

File Logical collection of data.

Flow chart Graphical representation of the operation of a program.

Gate A logic circuit having one or more inputs and a single output. For example, NAND gate.

Half duplex A two-way communication system that permits communication in one direction at a time.

Hardware The physical parts or electronic circuitry of a computer system.

Hexadecimal Base 16 numbering system. In hexadecimal notation, numbers are represented by the digits 0–9 and the characters A–F. For example, decimal number 165 is represented as A5.

High-level language Programming language in which each instruction or statement corresponds to several machine code instructions. Some high-level languages are BASIC, FORTRAN, C, PASCAL and so on.

Input device An external device connected to the input port of a computer. For example, a keyboard is an input device.

Input port Part of a computer that allows external signals to be passed into it. Microcomputer input ports are usually 8 bits wide.

I/O Short for Input Output.

Input/Output The hardware within the computer that connects the computer to external peripherals and devices.

Instruction cycle The process of fetching an instruction from memory and executing it.

Instruction set The complete set of instructions of a microprocessor or a microcomputer.

Interface To interconnect a computer to external devices and circuits.

Interrupt An external or internal event that suspends the normal program flow within a computer and causes entry into a special interrupt program (also called the interrupt service routine). For example, an external interrupt could be generated when a button is pressed. An internal interrupt could be generated when a timer reaches a certain value.

Interrupt vector Reserved memory locations where a program jumps when an interrupt is detected.

ISR Interrupt Service Routine. A program that is entered when an external or an internal interrupt occurrs. Interrupt service routines are usually high-priority routines.

K Multiplier for 1024. For example, 1 Kbyte is 1024 bytes.

Language A prescribed set of characters and symbols which is used to convey a program to a computer.

LCD Liquid Crystal Display. A low-powered display that operates on the principle of reflecting incident light. An LCD does not itself emit light. There are many varieties of LCDs. For example, numeric, alphanumeric, or graphical.

LED Light Emitting Diode. A semiconductor device that emits a light when a current is passed in the forward direction. There are many colours of LEDs. For example, red, yellow, green, and white.

Level triggered Circuit action initiated by the presence of a signal.

Logic levels Voltage levels representing the two logical states (0 and 1) of a digital signal. Logic HIGH is also called state 1 and logic LOW is called state 0.

LSD Least Significant Digit. The right-most digit. For example, the LSD of number 123 is 3.

Machine code Lowest level in which programs are written. Machine code is usually written in hexadecimal.

Microcomputer General-purpose computer using a microprocessor as the CPU. A microcomputer consists of a microprocessor, memory, and input/output.

Microprocessor A single large-scale integrated circuit which performs the functions of a CPU.

Mnemonic A programming shorthand using letters, numbers, and symbols adopted by each manufacturer to represent the instruction set of a microprocessor.

MSD Most Significant Digit. The left-most digit of a number. For example, the MSD of number 123 is 1.

Nibble A group of 4 binary bits.

NMOS Negative channel Metal Oxide Semiconductor. A device based on an n-channel field-effect transistor cell.

Non-volatile memory A semiconductor memory type that holds data even if power has been disconnected.

Octal Representation of numbers in base 8.

Op-code Operation Code. That part of an instruction which specifies the function to be performed.

Output device An external device connected to the output port of a computer. For example, a printer is an output device.

Output port Part of a computer that allows electrical signals to pass outside it. Microcomputer output ports are usually 8 bits wide.

Parity A binary digit added to the end of an array of bits to make the sum of all ones either odd or even. Parity is a method of checking the accuracy of transmitted or received binary data.

PDL Program Description Language. Representation of the control and data flow in a program using simple English-like sentences.

PEROM Flash Programmable and Erasable Memory. This type of memory can be erased and reprogrammed using electrical signals only, i.e. there is no need to use an ultraviolet light source to erase the memory.

Port An electrical logic circuit that is a signal input or output access point of a computer.

Programmed I/O The control of data flow in and out of a computer under software control.

PROM Programmable Read Only Memory. A type of semiconductor memory which can be programmed by the user using a special piece of equipment called a PROM programmer (or PROM blower).

Pull-up resistor A resistor connected to the output of an open collector (or open drain) transistor of a gate in order to load the output.

RAM Random Access Memory, also called read/write memory. Data in RAM is said to be volatile and it is present only as long as the chips have power supplied to them. When the power is cut off, all information disappears.

Register A storage element in a computer. A register is usually 8 bits wide in most microprocessors and microcomputers.

ROM Read Only Memory. A type of semiconductor memory that is read only.

RS232 An internationally recognized specification for serial data transfer between two devices.

Serial Information transfer on a single wire where each bit is transferred sequentially with a time delay in between.

Software Program.

Start bit The first bit sent in a serial communication. There is only one start bit in a frame of serial communication.

Stop bit The last bit sent in a serial communication. There can be one or two stop bits per frame of a serial communication.

Syntax The rules governing the structure of a programming language.

Transducer A device that converts a measurable quantity into an electronic signal. For example, a temperature transducer gives out an electrical signal which may be proportional to the temperature.

TTL Transistor Transistor Logic. A kind of bipolar digital circuit.

UART Universal Asynchronous Receiver Transmitter. This is a semiconductor chip that converts parallel data into serial form and serial data into parallel form. A UART is used in RS232 type serial communication.

USART Synchronous version of UART.

UV Ultraviolet light. Used to erase EPROM memories.

VDU Visual Display Unit.

Word A group of 16 binary digits.

Index

ADC0804, 132,137,138
Analogue to digital converter, 132
Architecture of AT89C2051, 4
ASCII, 147,167

BASIC, 16
Baud rate, 147
Bit, 16
Binary counter, 29
Buzzer, 85,86
Byte, 4

Calculator program, 161
C programming language, 13,15
C51, 15, 16
Counter, 10, 50
Compiler, 13, 15
CPU, 2, 12
CTS, 149
CU, 2

Data type, 16
Data memory, 2
Data width, 147
Digital thermometer, 119
Do-enddo, 24
DS1620, 109
DSR, 149
DTR, 149

Electronic siren, 95
End, 22
Electronic organ, 101
Enum, 17
EPROM, 1,3, 4
External interrupt, 10,12
Event counter, 75

Float, 18

Hexadecimal display, 46
High current buzzer, 87

IE0, 10
IE1, 10
If-then-else, 24
Interrupt, 11, 20
Interrupt number, 12, 20
Interrupt service routine, 20
Interrupt source, 12
INT0, 6, 78, 79, 80, 83
INT1, 7
ISR, 20
IT0, 10
IT1, 10

LED, 29, 57
LED dice, 38
LM35DZ, 132, 140
LSD, 50, 51

MAX232, 150, 152
Memory model, 19
Minimum configuration, 12
MOSFET, 89, 90, 92, 93, 94, 101, 131
MSD, 50, 51

Parity, 147
PASCAL, 15
PDL, 22, 29
PEROM, 2, 3, 4
Piezo sounder, 85
Pin configuration, 4
Program memory, 2
PSEN, 10

178 Index

RAM, 1, 3, 4
Repeat-until, 25
ROM, 1
RS232, 147, 148, 149, 150, 151, 155
RS232 connector, 148
RS232 signal level, 149
RST, 5
RTD, 107
RTS, 149
RXD, 150

Sbit, 17, 18
Serial port interrupt, 12
Sequencing, 24
Seven segment display, 57
Sfr, 17, 19
Sfr16, 17, 19
Signed char, 16
Signed int, 17
Signed long, 17
Signed short, 17
Sounder, 85
Start, 22
Start-end, 22
Stop bit, 147

TCON, 10
Temperature sensor, 107

TF0, 10
TF1, 10
Thermistor, 107
Thermocouple, 107
THIGH, 109, 111, 125, 131
TIL311, 53
Timer, 10
Timer interrupt, 12, 90
TLOW, 109
TMOD, 10
Transducer, 85
TR0, 10
TR1, 10
TSM5034, 63, 64, 69, 75, 83
TXD, 150, 152

Unsigned char, 16
Unsigned int, 18
Unsigned long, 17,18
Unsigned short, 17,18

VN66AFD, 89

WR, 8

XTAL, 6